No 1 Balloon Centre, Kidbrooke

Peter H Sydenham

Previous Midhurst WW2 Memoirs books.

Midhurst WW2 Memoirs: An Evacuee's Story. 2018.
Memoirs WW2 Midhurst: 1. A Place Close to My Heart. 2018.
Midhurst WW2 Memoirs: 2. 'Evil' Rising: 'Good' Awakening. 2020.
Midhurst WW2 Memoirs: 3. Lowest Times (1940-1943). 2023.
Midhurst WW2 Memoirs: 4. Hard Ending (1944-1950). 2025.

from Red Robin Publishing www.midhurstmemoirs.com

(Note as of 29 August 2025: This is the corrected version.)

Frontispiece: Recruitment poster. Wikimedia

No 1 Balloon Centre, Kidbrooke

by

Peter H Sydenham

Front cover: *Balloons protecting Buckingham Palace*

Red Robin Publishing

First Published 2024 by
Red Robin Publishing Pty Ltd
49/5 Mt Barker Road
Urrbrae
South Australia 5064
Tel: + 61404083339
Email: sydenham@senet.com.au

No 1 Balloon Centre, Kidbrooke

ISBN 978-0-6450071-1-4
Printed-On-Demand by Ingrams

Disclaimer
The publisher of this book has made every effort to ensure the
accuracy of the information contained in it but such accuracy cannot
be guaranteed. If errors have nevertheless crept in, the publisher
apologises, but cannot accept liability for loss of inconvenience
resulting from inaccurate information.

Dedication

This account is dedicated to the serving members
of RAF Squadrons 901, 902, 903 and 901/903
who gave their lives in
active service at the
No 1 Balloon Centre, RAF Kidbrooke, London
during WW2.

Preface

In my teens I sort of knew my father, Cpl Henry Sydenham, served as a Balloon Operator in WW2. He rarely talked about it. One day in 1951 he told me he was the only person of the ten people in his hut to survive a direct hit by a V1 bomb at the No 1 Balloon Centre, Kidbrooke. It took me a while to fill in the backstory.

As time passed, I was often asked by my children, grandchildren, and others: "What did your dad (grandad) do in WW2?". One of them gave me a well-constructed one-page essay as the result of our interview. That led to me taking up Henry's situation seriously.

In 1938, land between the current Thomas Tallis school and Kidbrooke railway station was the location of the RAF No. 1 Balloon Centre. During WW2 it flew barrage balloons to defend targets from German bombing.

RAF Balloon Command had twenty Centres that flew thousands of balloons in the British Isles and abroad. No1 Centre was the first to come into operation, with its Squadrons 901, 902 and 903. Personnel and support were needed for round the clock operation in all weathers.

It was a dangerous home-front service. Around 50, of the Balloon Command's 500 servicemen and women who lost their lives, died serving in this Centre.

Along my path to placing memorials at the site, its historic importance was recognised by the grant of a *Heritage Award* of the Royal Aeronautical Society.

Described here are the creation of the Centre, its facilities, record books and its post-war use. This book gives insight into life at the Centre. More detail, across national WW2 balloon use, is available in the numerous articles posted on www.bbrclub.org.

Peter Sydenham
10 January 2024

Contents

Chapter 1. Emergent Years of Balloons in War use

It was the day when Britain was expected to declare war on Germany; that is, on 3 September 1939.

Jonah Barrington was looking out from his workplace, a primitive hut in which was installed privately owned radio receiving station equipment. *Radio Towers,* near Leatherhead, was some 15 miles to the northeast of London. Jonah was a leading British journalist who had been directed to listen for any breaking news he could get from foreign radio broadcasts to use in his employer's newspaper.

In his field of view from that hut is a remarkable sight. Jonah wrote:

"And then the B.B.C. announcement - Chamberlain to speak at 11 a.m. So, it had come! I walked across the field and stared out across the fifteen miles of clear air to London - the capital which, in a few hours' time might be burning from end to end. And then my heart turned over! For there, in the sky, motionless and noiseless above the city, and outlining the city's contours, was what appeared to be a gigantic oval field of white, shimmering thistledown . . . Barrage balloons, [like in Fig. 1.1], - hundreds of them - glinting in the morning sun! The sight, so long expected and yet so utterly unexpected, brought a lump to the throat. A great city at bay! Defenceless, probably, down below, yet so gaily and gallantly defenceless . . .
Barrage balloons - pennons of modern warfare."
(Barrington, Jonah. *And Master of None,* Walter Edwards, London, 1948.)

Fig. 1.1 Balloons flying around London.

Claims of the earliest balloons suspended in the air, go back hundreds of years. A very light container was filled with heated air – the hot air balloon.

For our purposes the first one of real interest was that made and flown by the Montgolfier brothers in France. The first non-tethered flight, carrying people, flew to some 3000ft (910m) over a flight path of 5.6 miles (9km). This was achieved in the Montegolfier balloon that took place on 21 November 1783. An etching of that balloon is shown in Fig. 1.2.

Fig. 1.2 Coloured etching of the Montgolfier hot air balloon.

The Montgolfier balloon ascends once the air inside the light envelope becomes hot enough. It had no fire inside so it slowly dropped back to the ground.

The earliest engineering Professional Society was the Royal Aeronautical Society, that had its first public meeting on 27 June 1866, in London.

The early hot air balloons had many practical engineering problems to solve. The severe limitation of that time, the need for continuous heat to keep the air hot, made them of little practical use but they did spur on flight in the air, especially for flight by heavier than air machines.

Whilst they seemed to have more promise the balloon method was kept alive and developing to be able to carry very heavy payloads by filling the balloon frame with a lighter-than-air gas. That was already possible in the Montgolfier time using hydrogen, known about in the late 18th century and could be generated economically. That first demonstration of the use of lighter-than-air gas was shown by Jacques Charles and Tiberius Cavallo who engineered, at first, a small one with 9 kg lift. The *History of Balloons* website tells us that their engineering was then limited by the making of enough gas; they had to pour a quarter of a ton of sulphuric acid onto a half of a ton of scrap iron for their demonstration. It worked well and flew 21 km over 45 minutes – until the gas diffused through the varnished silk covering the delicate and light frame. Just a few days after the Montgolfier hot air balloon flew, these brothers made a 2h 5min manned flight, of 36km at 550m altitude, on 1st December 1783. It had some degree of altitude control.

Despite its high flammability hydrogen gas was used, being the only light gas that could be made economically. Helium was not available in large quantities until the early 20th century when it was first found emanating naturally from uranium ore.

So successful was the hydrogen filled balloon that the aerospace industry kept up their design for commercial and warfare uses until two major disasters took place of the British R101 (4 October 1930) and then the German Hindenburg (6 May 1937). Hydrogen filled airships were too dangerous and uncertain for human passenger application, but were workable for barrage balloon use where fire was an acceptable risk in air defence.

A balloon burning does not explode with force but simply burns away the gas bag material. The other major, largely unpreventable weakness was that lightning from weather storms could also set them on fire. British skies exhibit lightning so often!

Fig. 1.3 British Observer Balloon of WW1.

Development of the balloon in military use, Fig, 1.3, is well explained in *Balloons at War* by Christopher. By the 1930s that technology was very advanced, but some inherent weaknesses were never overcome. In military use they were too easily shot down. At first, ammunition from an airplane, or anti-aircraft (AA) gun, was ineffective for the bullets would simply pass through the covering without igniting the hydrogen; leaving only small holes that leaked the gas a little. It was soon realised that addition of tracer bullets in the gun bullet stream would easily set it on fire for those rounds heat up from a small pyrotechnic charge within them.

A balloon is basically a light bag filled with gas. The shape of balloon used in Britain during WW2 had been developed from experience over the period 1917 to the late 1930s. They had begun life during the First World War where the first barrage of balloons was used as 'aprons' in front of key sites needing protection from bombing.

The barrage is not merely a deterrent because of its presence over potential enemy targets. It is lethal also. The wires are tough enough that they cannot be cut by a plane's propeller. This means that raiders must avoid the wire or suffer damage sufficient to make them crash. It also has great psychological benefit; those living and working inside a balloon barrage area feel safer!

The approximate size of a WW2 British barrage balloon, Fig 1.4, is 64ft (19.5m) long and 31ft (9.5m) across. Its total weight is 600 pounds (272kg), Fig. 1.4.

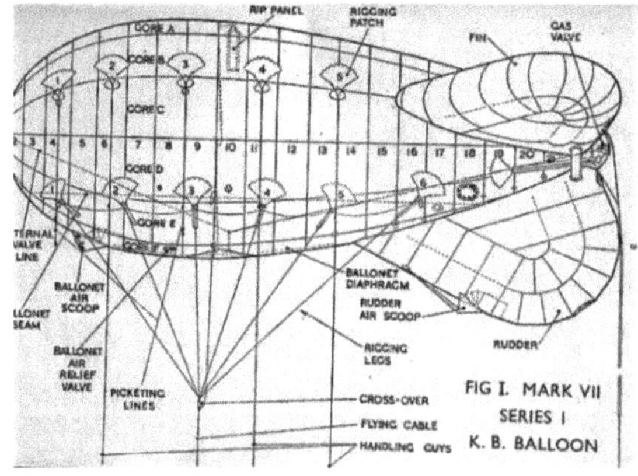

The first crews used only men; 16 were needed per balloon for all day and night operation. The job needed good physical strength. After a period into the war more men were needed to be released for active fighting duties.

1.4. Parts of a British Mk VII Balloon.

WAAFs (Women's Auxiliary Air Force) were trialled and rapidly accepted into this role, in which case a crew, of 20 or so, was used to operate a balloon all around the clock.

As well as the balloon operating crews there was need for around 50% more personnel to support the crews – in other roles. The balloon – see Fig. 1.4, flies with the rounded, smooth end, pointing into the wind. At the back, two top horizontal protrusions (ears) are stabilisers that keep the balloon from spinning. The bottom appendage (nose) is the rudder that keeps the balloon from oscillating from side to side. Scoops take air into these appendages to keep them inflated.

Picketing guy ropes are attached around the sides for holding the balloon down to fixed ground point anchors. Handling ropes facilitate manoeuvring it into position by the crew.

Safety mechanisms are provided - overpressure relief valves and a rip cord that rapidly releases hydrogen to obtain a very rapid descent. On the underside are the ropes fixed on the surface that are gathered to a single point – the cross over - where it hooks onto the steel flying cable from the winch truck. A pipe runs up this bundle to the hydrogen intake port on the balloon; it is used to fill, or top up, the hydrogen content.

A balloon operation begins with the fully deflated balloon fabric case resting on the tray at the rear of the special, purpose-built winch truck. The balloon is let up with hydrogen from the gas storage bottles and at first is held at 40ft to allow the air intakes to work, filling out the shapes. Ten of the 30 or more gas, cylinders on a trailer can be joined to a single point manifold to speed up inflation.

The winch operator can see the balloon and cable as it is let out and is seated inside a steel mesh covered position for protection. Tethering cables are most dangerous as they whip lash around after they break.

In gales, as balloons are blown up and down by violent gusts, terrific stresses are put on the ground equipment, and the cable-holding patches on the balloon surface. Winch trucks have been dragged by their balloon into a nearby dock. A winch truck was dragged to the cliff edge, where it poised precariously over a drop of 200ft. with a row of cottages immediately below. On one occasion a balloon took a steep dive during which the loosened cable wrapped around the bumper bar of a car; that was pulled off the ground. The passengers were rather taken by surprise!

Winch trucks, with their gear, each weigh five tons (5080 Kg). Although the barrages are mobile, they carry an enormous lot of other baggage. It took five special trains to transport a complete barrage to Scotland, with its lorries and winches going by road. Creating a controlled and robust system of hundreds of balloons took some time to put in place.

Fig. 1.5 Life was easier at times.

Looking at Fig. 1.5, whilst the lot of a balloon operator could be very tough, there were pleasant times when crews could relax, even with their families.

Chapter 2 Birth of Balloon Command

The Badge and motto of Balloon Command are given in Figure 2.1. The motto "Vi Et Ictu" means "By Force and Impact".

Figure 2.1 Balloon Command Badge.

A short general history encapsulating WW2 RAF Kidbrooke is provided on the BBRClub website.

With WW1 at an end the work on defensive balloons collapsed; a research establishment was maintained. Between 1928 and 1938 the Air Defence Command kept re-presenting the fact that it was necessary to be able to organise and maintain a barrage in any future war. Air staff decided in 1936 to establish a London barrage, with the provinces left unprotected until more experience was gained.

In 1936 the Committee of Imperial Defence approved the idea of 450 barrage balloons being set up to defend London. This initiated an eventual national balloon defence organisation. It was to be set up by The Auxiliary Air Force i.e., manned by non-regular, part time, volunteer staff.

No1 Balloon Training Unit was formed at RAF Cardington on 9 January 1937. On 17th March 1937 the first Balloon Barrage, Group No 30, was formed; that in which Kidbrooke sits. It was commanded by a retired RAF officer Air Commodore J.G Hearon CB, CBE, DS.

As war clouds gathered over Europe the fledgling RAF Balloon Command was formed in 1937 to prepare for use and control of all UK-based barrage balloons. The active service for Balloon Command was formed at RAF Stanmore in Middlesex on November 1st 1938 under the control of Fighter Command. At its head was AVM O.T. Boyd CB, OBE, MC with the title of Air Officer Commanding.

In 1939 it set up 47 Barrage Balloon squadrons, each typically having 5 balloons. They were given the squadron numbers 901-947 inclusively based on a county wide affiliation. (Later, more Squadrons were formed.) It was highly necessary that these Commands operated closely, for Allied aircraft had to know where and when balloons were, or should be, in operation at any given time.

At the outbreak of war in September 1939 the idea of a total of 1450 balloons was approved but there were only 624 in existence. Balloon production was 212 in September 1939 and fell to 148 in October. Production figures trebled as the months went by and by May 1940 the pre-war target was achieved, only to be faced with an imminent German invasion of France with Britain potentially next. It was at this time that waterborne balloons became to be used at ports and harbours. They were tethered to floating barges.

By 1939 improved designs had been approved along with a method of flying balloons so that dive bombing by German planes was frustrated thus reducing their capacity to bomb accurately. At the same time, it kept the enemy at heights that kept them within the range of anti-aircraft fire and could be better traced with searchlights needed for more accurate shooting by them.

The organisation and leadership of Balloon Command are listed at www.nevingtonwarmuseum.com/united-kingdom. In 1937 the first four barrage balloon depots for the defence of London were planned at as No.1 Kidbrooke; No.2 Hook; No.3 Stanmore; No.4 Chigwell.

Balloon training was carried out at the No.1 Balloon Training Unit at Cardington, Fig. 2.2, and No.2 at Larkhill. Training included three months driving and three months on balloon handling and wire rope splicing.

Fig. 2.2 Balloons on their winches ready to "let up" at Cardington.

At the end of the course, trainees took part in a 50mile run in convoy, to inflate a balloon and fly it. By August 1938 it was decided to open the trade of Balloon Operator to man new balloon depots being formed around the country.

The principal objective for Balloon Command was to maintain a sufficiently dangerous 'forest' of hanging cables which would make it very difficult for enemy bombers to make accurate attacks on the areas below. To do this it was obviously necessary to extend the 'barrage' over a wide area and the balloon squadron used a variety of mooring areas which included major cities, ports, naval bases, and industrial centres. Mooring points included both fixed types and winch lorries.

As a result of expanding the balloon barrage to include other cities outside London, direct control changed from Fighter Command to a command of its own. Balloon Command was, therefore, formed on 1 November 1938 under the command of Air Vice Marshall Boyd.

'Balloon Command' divided Britain on a geographical basis, into 'Balloon Groups' and each one of those was, in turn, subdivided into 'Balloon Centres' staffed by 'Squadrons' of Balloon Operators who were each responsible for several 'Flights', who flew their individual balloons on nominated 'Sites'.

Each balloon Centre was a depot for the supply and repair of equipment, along with balloon maintenance. The balloons were not always flown from fixed static sites; balloons were also flown from barges, drifters and mobile lorries.

To get the bases up to strength, a recruitment drive took place. One of its posters is given in Fig. 2.3.

The life of Balloon Command was relatively short. The Balloon Training Unit closed down in 1943. By then it had trained over 10,000 RAF and WAAF balloon operators and some 12,000 operator drivers.

Fig 2.3 Recruitment poster.

The key appointments that got Barrage Ballooning all going as one, were first its leadership, and then its place within the RAF structure. Newspapers reported developments of these two aspects:

"LONDON BALLOON BARRAGE; TONBRIDGE OFFICER'S APPOINTMENT

Major J. S. Wheelwright. D.S.C., of Three Elm-lane, Tonbridge, has been appointed to command No. 901 (County of London) - Squadron of the new balloon barrage for the defence of London. He is one of the first six commanding officers to be appointed and, with the others., has been granted a commission as Squadron Leader in the Auxiliary Air Force. Major Wheelwright is a balloon designer and a qualified airship and balloon pilot. During the war he served in the R.N.AS. and R.A.P. and gained the D.S.C. in 1917 as a Flight-Commander for services on patrol duties and submarine searching in home waters. The nearest point to Kent in this defensive scheme is Kidbrooke, and there they require 1,500 men as recruits for the Auxiliary Air Force. All the officers have been recruited. The men required should be between the ages of 38 and 50 and preferably should have served in the Forces before, though that is not strictly necessary. They should have some knowledge of motor driving or mechanical knowledge, or they could be trained riggers. Three squadrons are stationed at Kidbrooke, where splendid accommodation has been provided for the men. The commanding officer is Air-Commodore Bowen. Any man wishing to enrol can report at the headquarters at Kidbrooke, or information can be obtained from any Territorial centre. Expenses for transport are provided."

Kent & Sussex Courier 27/5/1938

The issue of ownership of this new Air Force entity came about after considerable debate:

"LONDON'S BALLOON BARRAGE SCHEME NO LONGER NOBODY'S CHILD "

General confusion, which has hampered development London's balloon barrage system, is to be dispelled by a much-needed co-ordination of command.   Youngest branch of the service, it had

11

become the nobody's child of five separate organisations. For instance, the Balloon Development Establishment for recruit training at Cardington (Bedford) has been under direct administration by the Air Ministry. The three balloon centres, at Kidbrooke, Hook and Chigwell, were grouped under No. 30 Balloon Barrage Group, which was administered, in turn, by the R.A.F. Fighter Command. And, to make confusion worse confounded, the ten-balloon squadrons were units of the Auxiliary Air Force. The Air Minister, Sir Kingsley Wood, told Parliament last week that the balloon defence system would be extended to 12

provincial centres. To meet this expansion and to unify its administration, the Air Ministry announce to-day the creation of a separate Balloon Command and the appointment of Air Vice-Marshal Owen Tudor Boyd, [Fig. 2.4], as Air Officer commanding all balloon organisations.

Fig 2.4 Air Vice-Marshal O T Boyd.

"One of the new Commander's first tasks may be to secure some degree of uniformity among officer London's ten balloon squadrons, not only in numbers but in relative ranks. Their disparity is well illustrated in the three County of London squadrons stationed at Kidbrooke, each with seven officers. Those attached to No. 901 Squadron comprise squadron leader, a flight lieutenant and five flying officers; No. 902 has squadron leader, flight lieutenant, four flying officers, and two acting pilot officers; No. 903 carries on with a squadron leader, two flight lieutenants, two Flying officers and two acting pilot officers."
22/11/1938 The Citizen

Their acceptance as part of the RAF was not all easy; there were those who felt that this was not a flying force so should not be said to be 'flying, but 'sailing'!

Elsewhere it has been said:

> "Eventually, this command consisted of 33,000 men and women. The amount of equipment and the number of personnel, however, tell only part of the story. Performance in combat is the main indicator of a weapon system's success, and the balloons received a thorough test during World War II.
> By the middle of 1940, there were 1,400 balloons, a third of them over the London area. By 1944 the number had risen to nearly 3,000. Later the barrage balloons were moved to combat the V-1 flying bomb. By 1944 the balloons were moved to make up a ring around south London to combat the V-1 menace with a fair degree of success - as many as 100 V-1s snagged themselves on the balloons' cables. It was not all plain sailing, however. Some of the balloons were struck by lightning while others were shot down - 50 were shot down in one day when they were set up round Dover."

Seeking to innumerate how dangerous balloon service was, a crude measure is to take the whole count of personnel of all roles in Balloon service (33,000) and compare these with the deaths now recorded (around 600); this is 2 deaths per 100. Losses in the D-Day invasion, counted the same way, were some 4400 deaths in the Allied Force of 156,000, that being a rate of 3 deaths per 100.

Clearly, operating barrage balloons was a major active home front undertaking, performed at high risk under fire, much of the time necessarily undertaken in the worst of weathers. It also demanded continuous round the clock service with little 'down' time.

The genesis of the No 1 Centre is covered in a BBRClub article on its website - *Formation of a Balloon Depot at Kidbrooke.* In brief Wing Commander Maude (Ret) announced, on 12 January 1937, that balloon training was underway and that four main depots needed to be created. Sites were needed to be acquisitioned, as would be the building of accommodation and support facilities. Those requirements were hard to find in the Greater London area but eventually Kidbrooke was the best choice. The already RAF owned No 1 Stores Depot, plus the purchase of an adjacent farm to the east, obtained the 40 acres needed. Formal establishment took place on 4 October 1937 under the command of the Air Commodore J B Bowen.

An advance party was in place on the land to ready it for operation. An important step was the approved provision, by NAAFI (Navy, Army and Air Force Institute) of 4 meals a day plus supper – but with a charge to each person of 2s 10d.

A plan of the site use is given in that BBRClub article; it was quite a bit different by the end of the war.

The operational birth times of the 901, 902 and 903 Squadrons, that formed the No 1 Balloon Centre at Kidbrooke, are recorded in their Operational Record Books (ORB) - that being on 16 May 1938.

They had much the same personnel strength, being an approximate total of RAF 67 and AAF 550 each, that is, approximately one tenth were embodied persons, the others being auxiliaries. Each of these squadrons flew around 45 balloons.

As the war progressed the men operating balloons were needed

in active service elsewhere. Women were trialled in their role and were rapidly accepted as being competent, but needing a larger crew. They became the major part of personnel of No 1 Balloon Centre.

Fig 2.5 WAAFs on parade with balloons.

Chapter 3. Use of RAF Kidbrooke Over its Life

RAF Kidbrooke was not an operational flying machine airfield and thus was less glamorous than many airfield bases ... but it was a major support site.

From 1938-1947 it was the RAF No1 Maintenance Unit. At the time it took in No1 Balloon Centre it had services and support for over a thousand vehicles and possibly workspaces for at least 2000 personnel, supporting the many balloon operations and including the already existing vehicle maintenance, aircraft storage, modification and ordnance repair. Balloon support came to the base with urgency in late 1938, with some specialised extra building work to help them all fit.

It seems reasonable to state that Balloon Centre 1 was initially shoehorned into available space and made use of many already existing support services. RAF Kidbrooke certainly was not as purpose built for balloons as were other contemporaneous balloon Centres, such as those at Stanmore and at Hook.

Obtaining detail about the Kidbrooke operation needs more research into the off-line records at the National Archives and the Imperial War Museum – and that needs people who live close to those record keeping places; and with time to donate!

Useful information is, however, now to hand that allows a first-cut picture to be painted as to the size and scope of RAF Kidbrooke with respect to the balloon use from 1937 until 1945, when Balloon Command was disbanded.

Contemporary photos of the site before the balloons came are rare. One photo found of RAF Kidbrooke for the period before WW2 is of a row of boilers installed in 1919! It is not shown here. It was taken for G N Haden and Sons Limited, heating engineers. (IMW image BL24694).

An interesting 3-page article on one aspect of pre-WW2 use is *Kidbrooke. The Temple of Truth,* by L G Callingham, Popular Flying, vol 1, no 8, November 1932, p439. It covers the testing laboratories there for petrol combustion, photography, electricity variables including testing thermionic valves, measuring devices of many kinds, doping of fabric, microscopic features, and RAF uniform wear and strength. Fig. 3.1.

Fig. 3.1 Appliances for examination of fabrics and cloth

Several local people have published research notes on the use of the RAF Kidbrooke site, over its life.

First there was an enquiry sent to the Ministry of Defence in January 2019 by John King, (Ref: AHB (RAF) 0014/2019):

'The site at Kidbrooke was in use by the Royal Flying Corps (precursor to the Royal Air Force) from mid-1917, when No 1 Stores Depot was established at the site. Stores at the depot included aircraft, vehicle spares and tools and electrical, wireless, photographic and explosives stores. The site continued use as a stores depot by the Air Ministry/Ministry of Defence until early in 1970. At various times the site also included a Wireless Transmitter (WT) Repair Section, a Gun Testing and Repair Section, a Non-metallic and Chemical Test Section, a WT Station and an Air Publications and Forms Store. The site at Kidbrooke was also used for varying periods to house No 1 Balloon Centre, No 4 Motorised Transportation (MT) Squadron, a Special Signals Unit, a School of Stores Accounting and Storekeeping, an

Equipment Officers Training School, and an RAF Movements School.'

David Wise, of Eltham, an allotment user, has researched this site over recent years and provided his understanding on the web site of the Kidbrooke Park Allotment Association; see www.kpaa.org.uk/history.htm. The Allotment has become an important part of this story. His text of 2000 is:

'The RAF presence at Kidbrooke started with the establishment of the Royal Flying Corps Stores Depot there in June 1917; later an Aircraft Storage Unit (ASU). It was retained between the wars and expanded. The RFC became the RAF on 1/1/1918. Kidbrooke was renamed No 1 Stores Depot on 1/3/1920, No 1 Equipment Depot on 2/2/1937 and the Repair depot called No 1 Maintenance Unit (No 1 MU) in 9/4/38. Its store's function was officially disbanded as a unit in its own right on 15/2/47 but continued as a satellite managed from elsewhere. It also housed some special units until the early 1960's. It grew in size. There were sites on both sides of the railway and both east and west of Kidbrooke station. Most of it has now been redeveloped for public and commercial use, but some warehouses NE of the station are in use as a store for Greenwich Maritime Museum, accessed from Nelson Mandela Road. The site NW of the station post-war became the RAF Movements School. 1956-1961 training RAF cargo-handling staff; and I remember the fuselage of a post 1950 Hastings transport aircraft there, Jan 1959 to mid-1961, which was clearly visible from passing trains. It also housed RAF No 4 Motor Transport pool until 1964. This site was later partly used for Thomas Tallis School, and partly as a Post Office Vehicle Depot then currently for warehousing.'

[Ed. The school, now with 1985 students in 2022, began its life there in 1971, but was completely rebuilt in 2011 as part of the *Building Schools for the Future* programme.]

The Kidbrooke site had also housed, between 1949-1953, the RAF Joint Services School for Linguistics. Fig. 3.2 is likely to have been a group there. A commemorative plaque to this RAF school was unveiled at Thomas Tallis School in 2008, see later Fig. 10.2. It was an Air Ministry initiative to increase the number of Russian speakers.

Fig. 3.2 Students at a Joint Services School for Linguistics.

'The main stores depot south of the railway had a railway siding accessed through a gate west of Kidbrooke station with its own shunter, diesel post-war, but originally steam - two steam locos which are known to have worked here are still in existence, one in a museum in Leeds and one active on the Bluebell Railway. There was also a 2 ft narrow-gauge railway within the site, including a bridge under Kidbrooke Park Road, subsequently used as a roadway when the site was redeveloped as the (now replaced) Ferrier estate. A 1917-vintage narrow-gauge tank engine named Kidbrooke is in service on the Yaxham Light Railway in Norfolk.

I am not sure which of the buildings they used, but during WW-2 Kidbrooke also housed a variety of units apart from its stores function. These included No 1 Balloon Depot which was parent to several Balloon squadrons numbered upwards from 901, operating barrage balloons from sites in and around London; also No 2 Installation Unit, which was responsible for constructing and repairing the chain of home radar station masts at many locations round the coast.

In WW-2, No 141 Gliding School for the Air Training Corps was on a different site altogether. It used the patch of ground, part of which survives as the Dursley Road ILEA Playing Fields. It was then much bigger, bordered by Broad Walk to the South,

Woolacombe Road to the West, Wricklemarsh Road to the North, and the Western wall of Brook Hospital to the East. Since the war, parts of the site have been built on, creating Dursley Road, Holburne Road and the associated side-roads south of Wricklemarsh Road, plus Corelli Road and Kidbrooke School. The gliding school was in operation October 1942 to December 1945.
I am still researching the history of the allotments site and if you have any useful information, or pictures, please forward these to me at kidbrookewaitinglist@gmail.com. David Wise KPAA."

Recorded memories of those who worked during the WW2 at RAF Kidbrooke are scarce. Alan Mann had some experience there as a mechanical fitter:

'In 1938, aged 12, I won a scholarship for the South East London Technical Institute (SELTI) enabling me to start a 3 year full time engineering course, which I completed early in 1941. My first job was at the Redwing Aircraft Co, Croydon, which I joined in the summer of 1941. I was employed helping to make self-sealing fuel tanks for Wellington bombers. After a few weeks at Redwings, I soon realised this was not the career I had in mind and approached the RAF at Kidbrooke to see what was available.
Kidbrooke was the home of the No.1 Maintenance Unit and No 1 Balloon Centre; but to me it was also the base for the RAF's Dance Band, the Skyrockets. Kidbrooke offered a seven-year mechanical engineering trade apprenticeship which I accepted, commencing the apprenticeship on the 17th of June 1941. Kidbrooke was also the home of the first Air Training Corps Gliding School (No. 141), headed by Squadron Leader Furlong, a competent gliding pilot. I had already joined the ATC shortly after it was formed in February 1941, whilst still at the SELTI. At Kidbrooke I initially worked with an ex-Rolls Royce engineer helping to repair Merlin engines and more mundane things like making trolleys containing the batteries required for starting

aircraft engines. After a couple of years, I became restless and wanted to join the RAF, as a pilot of course. After much persistence on my part, I wrote to de Havilland at Hatfield and was interviewed at their Stag Lane engine division at Edgware. I was asked questions about my previous schooling, works experience and why I had chosen de Havilland to further my career, this last part was easy! After what seemed like a lifetime, I learnt that I had been successful and was offered a position as a trade apprentice, engine fitter and tester, which I gladly accepted.'
http://www.memoriesofwar.org.uk/documents/Some_of_my_tee nage_memories.pdf. Alan Mann, 2006.

In Balloon Command it was soon decided to train WAAF personnel, as well as male operators; to release men for other forms of active duty. An ORB for the Centre recorded the change-over to WAAF crews on 28 February 1942. The personnel strength data from the ORBs (Operations Record Book) for 1943 shows that the use of WAAFs increased markedly during that year.

There were those who scorned the idea; but the WAAF's set-to and showed they were more than ready to meet the challenge.

The article, *The Development of Balloon Command* on www.bbrclub.org gives detail on staffing:

'The Balloons were flown by two corporals and ten airmen on a round the clock shift basis. In 1940 the idea was floated that balloons could be operated by WAAF's to release men for active service in other areas. This idea was hotly contested by AOC Balloon Command and his WAAF Staff Officer. 20 WAAF Balloon Fabric Operators were trained in London in April 1941. In May 1941 the first batch of WAAF volunteers were posted to a 10-week training course at a balloon centre. Initially, the powers that be, working on the basis that 10 male balloon operators could only be replaced by at least 20 female balloon operators, began to substitute women for men in the squadrons. The women rose to the challenge and showed that 14 women were quite capable of replacing 10 men and this was eventually settled as the correct figure. By December 1942 10,000 men had been replaced by some 15,700 WAAF balloon operators. For those squadrons who went abroad to defend vital installations, the operators were all male.

The balloon barrage was always a very risky area to fly aircraft of any sort in. Friday, 24th May 1940 was the day on which the balloon barrage was to claim its first victim, unfortunately one of our own RAF planes! This first aircraft downed by the balloon barrage was an RAF Hampden bomber, Fig. 3.3, which hit the barrage over Coventry in the daytime and landed in the cricket ground.

Fig. 3.3 RAF Hampden bomber.

Another Hampden collided with the barrage on 4th June 1940, at Shotley, 2 miles NW of Harwich - this is the first balloon casualty to be recorded by Home Security. On 13th June, a third Hampden hit the barrage at Harwich and crashed in the dock area at Felixstowe. Only one man survived from these 3 crashes.

The first enemy aircraft brought down by the balloon barrage was on 13th September 1940. This was a nighttime incident and this first German loss, reliably attributed to the Balloon Barrage, was the loss of a Heinkel He 111, claimed by a mobile unit of 966 Squadron on station at Belle View Park, Monmouthshire. The plane was returning from a raid on Merseyside when it

struck the cable and plunged into a built-up area of the above district. On the ground two children were killed. Three of the aircraft's crew were killed, the pilot managed to bail out in time. The aircraft was destroyed.
Flying of balloon barrages was ended in the United Kingdom in autumn 1944. This led to the disbanding of Balloon Command in February 1945.'

Some interesting, but with insufficient image fidelity, history has been archived by the Pathe Film project, see www.britishpathe.com. Searching with 'Kidbrooke' finds two historic films of Balloon Command events at Kidbrooke. These short films are expensive to purchase but can be viewed with streaming for free - with some constraints!

The earliest film, in September 1938, shows a visit by Air Minister Sir Kingsley Wood, Fig. 3.4.

Fig. 3.4 Air Minister sees barrage balloons. 1938.

Its précis states it shows balloons, a winch truck and cylinder trailer, plus many airmen in a relaxed mood. This was made to assist the 1938 major recruitment campaign for 5000 men (1500 for Kidbrooke) between 25-50 years of age (see later) to operate the hundreds of

balloons to be sailed around London by a year's time when war was declared. It stated the visit was to the "1st Barrage Balloon Squadron" (that is presumably, Squadron 901).

Fig. 3.5 RAF and WAAF parade. Kidbrooke 1941.

A second Pathe film, Fig. 3.5, of 1941 shows a parade of men and women Royal Air Force members. Some 80 men and 40 women are parading with an RAF band. It is not stated who is parading but the number and mix of men and women would suggest they are the balloon operating crews of Balloon Squadrons 901, 902 or 903, who were stationed there by then.

Another, amusing, photographic angle on barrage balloons is found in *The Day the Balloon Went Up,* Dad's Army: Series 3:8, written by Jimmy Perry and David Croft, and screened by the BBC. The *Midhurst and Petworth Times* newspaper of 14 October 1944 carried a short article on a barrage balloon that came into that District trailing its cable, demolishing chimney pots and cabbages. Perhaps this incident was the one that prompted this episode of "Dad's Army". However, numerous barrage balloons went astray causing damage along their way.

From 1938–1945 RAF Kidbrooke operated some 140 barrage balloons at any given time. That needed around 1000 vehicles of several kinds and some 2000 service personnel to operate and maintain - see list in BBRClub Newsletter Spring 2014 and several Peter Garwood's accounts on www.bbrclub.org .

As already said, there are few photos of RAF Kidbrooke activity but that, given as Fig. 3.6, has been recorded on the BBRClub website. It shows the extent of a Squadron with its five balloons and support trucks in place. It is more likely "B" Flight of Squadron 902.

Fig. 3.6 A Balloon Flight from RAF Kidbrooke.

Fig. 3.7 This balloon is not behaving well in front of visiting VIPs.

Fig. 3.7 shows that balloons could be difficult, especially when the wind does not cooperate, and people are watching. It also shows why at least ten men are needed in a crew. More information is to be seen on www.Thamesfacingeast.com. On January 12, 2014, it stated:

'The RAF station was developed in 1916 on both sides of the railway line close to Kidbrooke Station. Opposite the station, the land, which is now Thomas Tallis School, was a prisoner of war camp. The RAF station was built mainly by the prisoners. The RAF had a presence in the area until after the Second World War and closed in 1965." ……………..
"The Air Ministry bought land that had been the Lower Kidbrooke farm now the Carnbrook Road area. This was used for making barrage balloons which defended London against low flying aeroplanes." ……………..
By 1941 the Luftwaffe was confident that they were ahead of the British in terms of radar developments. Yet, by the end of the war

this lead was lost and the allies had superior capabilities. The Oboe bomb sight was made in Kidbrooke and was a key radar device that ensured the Allies' supremacy.'

Oboe was a position location beam system that gave information to an airplane as to where it was. It needed a lot of the latest electronic technology and could only service one aircraft at a time. It did, however, allow a pathfinder aircraft to locate a bombing target and drop a marker flare; that made their run far more effective. In its first deployment in December 1942, the large, and very heavy viewing system could only be used on the ground; not installed in an aircraft, Fig. 3.8, possibly a mobile one used after the invasion of D-Day 1944. Today a far superior capability is built into a wristwatch – GPS!

Fig. 3.8 Oboe console.

Radio signals, X, Y and Z - indicating the deviation error size, were sent to the plane by morse code to tell the pilot which way to turn to keep it on the needed, slightly curved, line to the target. S was sent to indicate the approach was starting.

The Kidbrooke Park Allotment Association adds more to the history, see www.kpaa.org.uk.

'Although the allotment land is owned by Greenwich Borough Council, the allotments are a self-managed site. The Kidbrooke Park Allotment Association was formed incorporating the existing Kidbrooke Park Allotment and Garden Society. .... The allotments first came into existence about 30 years ago, prior to which they had during the Second World War been known as RAF Kidbrooke, where 901 Squadron [Added: also 902, 903] were stationed. 901 Squadron was a barrage balloon squadron, and a barrage balloon was flown above the area up until 1944 to deter low flying enemy aircraft. It is also believed that the area, including where the existing Thomas Tallis School is situated, was also an airfield for flying model gliders after the Second World War'

Neil Rhind, a local history expert, was consulted to verify these suggestions but his area of interest, adjacent Blackheath, did not extend to include Kidbrooke.

A report, in the unexpected location of the *The Mercury* newspaper (Hobart, Tasmania, Australia), on the 1st March 1940, described the pros and cons of using balloons in the defence from aircraft:

'THE BALLOON BARRAGE
London's Protection Against Air Raiders

FORTY THOUSAND men, 10 times as many as a year ago, now man Britain's barrage balloons, and their numbers are still increasing as more and more barrages are sent up. Today the barrages guard vulnerable strategic points as well as the principal cities.

When, after the Nazi air raid on the Firth of Forth, (on 16 October 1939) it was decided to put up balloons to guard the Forth Bridge, the barrage was transported and put up again in 24 hours (states the London "Daily Mail"). [Added: The facts are that the bomber was after a ship in the estuary, not the Bridge].

It was explained by an Air Ministry official In London that the barrage is not merely deterrent because of its presence over potential enemy targets. It is also lethal. The wires must be so tough that they cannot be cut by a 'plane's propeller. This means that raiders must avoid the wire or suffer damage sufficient to make them crash.

The height to which the balloon will carry the wire comes second. The balloons are intended only to stop low-flying and dive-bombing tactics and to keep the enemy at a height which makes him the best target for anti-aircraft fire.

In no sense are the balloons supposed to be an "umbrella" to prevent enemy machines from coming over their targets at all.

"That." said the official, "Is a point worth remembering by those who come from country districts into cities to live 'In safety' under the shadow of the balloons."

Since the inauguration of the barrage many lessons have been learnt, including the need for many more meteorological reports. The balloon acts as a perfect lightning conductor and is apt to be vulnerable to thunderstorms.

If a balloon breaks away, the amount of damage it causes depends on the length of wire it trails. Carrying several thousands of feet of cable, it goes hopping along, alternately sinking until the ground takes the weight of some of the cable and then rising and taking the cable into the air with it.

These lorries with their gear each weigh five tons. Although the barrages are to some extent mobile, they carry an enormous lot of baggage, and it took five special trains to transport a barrage to Scotland, though the lorries and winches went by road.

Among the points mentioned by the official was that the balloons each cost 6s/8d a day to keep filled. That is, they lose a thirtieth of their hydrogen daily and a full balloon takes about £10 worth of hydrogen.

Although the prevailing winds blow to Germany and not from Germany; two Nazi balloons have been salvaged in England. One of these balloons recently came down after a 20-mile journey across the Shetlands. It had presumably made a 500-miles journey from Germany. Printed in German on it were instructions for it to be sent to Heidelberg.'

The Forth Bridge balloons referred to above are shown in a 23-minute propaganda film, giving operational detail, by the Royal Mail Group of the GPO - '*Squadron 992 (1940)*'. It was made after a German bomber attempted to bomb the Forth railway bridge area in 1939. This, general history, record filming unit was taken over by the Ministry of Information soon after this film was shot. It is available for viewing on www.screenonline.org.uk/film/id/1364857/index.html, but there it is only available to registered persons in the UK!   It is also on YouTube, and is included in the BFI DVD compilation '*If War Should Come: The GPO Film Unit Collection Volume 3*, available for purchase on Amazon.com.

If you don't mind the watermark on frames, it can be viewed free on-line at:

http://www.iwm.org.uk/collections/item/object/1060005895

Fig. 3.9
Squadron 922
in convoy to
Scotland, 1939.

From this film, a screen shot shows the convoy, Fig. 3.9, is probably the best WW2 UK Barrage Balloon activity film available. It shows men from the newly formed 992 Squadron being trained as a mobile unit and then moving up the country, needing 24 hours to commence their active service of protecting the Forth Bridge. The journey was around 450 miles (700km).

The convoy for a full squadron of hundreds of vehicles was very long and could only move at night at a very slow pace, due to blackout requirements of 20mph maximum and convoy dynamics that invariably need latter vehicles to move faster at times than the first. The film was possibly not made of the actual situation; it may not have been at RAF Kidbrooke, but their functions would have been the same. Again, the facts seemed to shift over time. To make it harder for historians to piece together the truth it will be seen that Squadron 992 is stated elsewhere to have been formed in Great Yarmouth in December 1943, yet the raid was in 1939 and they were formed at that time! The film was made in 1939 and is about Squadron 992.

Another film in the BFI collection shows girls 'learning the ropes' as they first trained to become Balloon Operators, they then being shown going into operation. It is called *Balloon Site 568 (Short 1942)* in the IWM film collection.

Operation Outward: Britain's WWII Balloon Attacks Against Nazi Germany is a short You Tube film about another use of hydrogen filled balloons as weapons.

In that British operation 99,142 balloons were let loose to float, from Britain and other eastward countries, across the European theatre into nazi held territories. Some carried trailing wires to entangle with objects, or to short-circuit overhead electrical power lines.

Others took incendiaries device in bags, set off with slow burning fuses. These could start forest, or buildings fires.

The first were set off in March 1942 to float toward Berlin. By August 1942 over 1000 per day were being launched, mainly by

Royal Navy Air Service RANS women. The end of this kind of use of balloons was soon after the D-Day landings in June 1944.

Today, some of the original RAF Kidbrooke site has been turned into a popular public garden allotment site. As we shall see later, that is fortunate for much of the rest of the site is now covered with modern buildings.

Chapter 4. Balloons Become an RAF Command

RAF Kidbrooke was the home of No. 30 (Balloon Barrage) Group. The overall Balloon Command went into active service use on November 1938 at RAF Stanmore Park in Middlesex. It consisted of its headquarters and four County groups. (Kidbrooke for London; Hook for Surrey; Stanmore for Middlesex; and Chigwell for Essex.) We will return to Stanmore later.

The locations for these are at the corner outer points of the Greater London area:

- Balloon Centre Kidbrooke. London. SE. Map ref: TQ408758.
- Balloon Centre Hook, Surrey. SW. Map ref: TQ174637.
- Balloon Centre Stanmore, Middlesex. NW. Map ref: TQ165921.
- Balloon Centre Chigwell, Essex. NE. Map ref: TQ4432944.

From 1938-1947 RAF Kidbrooke was also home of RAF No 1 Maintenance Unit. In that period, it also became the base for the start-up of operations for Balloon Command's first active service in Squadrons of the 900 series, here being the location at Balloon Centre No 1, Kidbrooke, London SE3 with its Squadrons 901, 902 and 903. Their balloons were run from HQs here and in the Abbey Wood area, maybe others.

An example of the colossus structure of the RAF, with the names of the senior officers, including those of Balloon Command (here was seen in the May 1944 issue, on its p96), are available in a published document called the *Air Force List* available from https://archive.org/stream/airforcelistmay1944grea#page/n5/mode/2up . During the war years this list was updated from 1-3 monthly intervals.

It is a free, but very large, download. In mid-1938 a major recruitment drive took place over Britain to find the many men, in the age bracket 25-50, needed to meet the large embodiment targets, see Fig. 4.1.

Fig. 4.1 Poster put up to attract voluntary enlistment into RAF Balloon Command, 1938.

A good start to delving into the life of any individual in the Air Force can be found in a person's two main military records - *Service Record* and *Service Medical Record*.

The process for obtaining a military *Service Record* for a person has become so much easier over the past decade. Start with this address: WW2 RAF Records - 4.2 M+ 1939-1945 WW2 Records

It is helpful if one can supply a service number and unit name, but they are not essential to get started.

Little is known about the service experiences of the estimated 2000 plus personnel of Balloon Centre No 1 at Kidbrooke; a few glimpses are available in the records.

One is that of Henry Sydenham, my father and one, may be two, of his colleagues who were all working together as *messengers*, at the Royal Exchange Assurance in Central London in 1938, Fig. 4.2. The messengers physically ran monetary documents to clients within the City of London, a once very necessary and honourable job that was fast becoming redundant in the late 1930, due to technology changes.

Fig. 4.2 Royal Exchange.

They must have gone to the same recruitment office in October 1938,

and at the same time, for they have almost consecutive RAF service numbers; Stanley Ross 840928; Bill Murray 840930; and Henry Sydenham 840931. They all started out in Squadron 902. Stanley Ross, Fig. 4.3, was the father of Vera Pullin, BBRClub Member No 440. Henry did not serve with them for he joined the RAF, not the Auxiliary Air Force.

Fig. 4.3 Stanley Ross, seated, with crew.

A fuller account of Henry's, Fig.4.4, story is included in the article on

www.bbrclub.org *840931 Cpl Henry Sydenham RAF No. 1 Balloon Centre (901, 902 Squadrons)*,

Fig. 4.4 Cpl Henry Sydenham, 1940 with his son Peter, the author.

Before war broke out balloon men usually first joined the Auxiliary Air Force, A.A.F; fitting in learning the ropes, so as to speak, part-time when living at home.

Henry decided to join up full time, the other two men known to have been with him, Stan and Bill. Went into the part time A.A.F. This arm of the RAF is described in a booklet published to assist recruitment in 1938. Its title is yet to be found. This included Barrage Balloon work. Relevant pages, 36-39 and 53, of the A.A.F. Recruitment Booklet appeared in the *BBRC Newsletter* of Christmas 2013.

When war broke out in September 1939, Stan and Bill were changed from AAF, being enlisted into full-time embodiment in the Balloon Command. Henry had gone into immediate embodiment in October 1938.

A good set of records of A.A.F airman Stan Ross in Squadron 902 are held by the Vera and Arthur Pullin family, BBRC Members 440 and 373 respectively. Vera is the daughter of Stan Ross.

The memories of Stanley Ross, no longer with us now, were dictated around 2007. They provide a good account of life on the base. His notes are to be found on the BBRClub website as article *Personal History of Stanley A Ross 902 Squadron.* They provide much detail on the life as a ballooner, but not much on the details of the Kidbrooke base.

The Summer 2011 BBRClub Newsletter contains copies of his "calling out" or "calling up" paper, Fig. 4.5, on 23 August 1939, his Release Book entries, and his Certificate of Discharge and Release Authorisation.

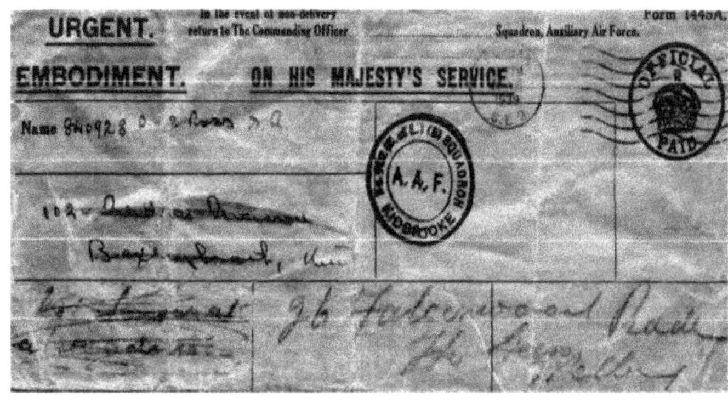

Fig. 4.5 Stan Ross's embodiment notice.

The Pullin family also has his pay book. Thank heavens some folks kept their family records!'

Non-Commissioned Rank progression through the RAF was:

Abbreviation	R.A.F. (and A.A.F?)	W.A.A.F. - post Jan 1940
AC2	Aircraftman 2nd Class	Aircraftwoman 2nd Class
AC1	Aircraftman 1st Class	Aircraftwoman 1st Class
LAC	Leading Aircraftman	Leading Aircraftwoman
CPL	Corporal	Corporal
SGT	Sergeant	Sergeant
F/SGT	Flight Sergeant	Flight Sergeant or Senior Sergeant
WO	Warrant Officer	Warrant Officer or Under Officer

A medal was awarded for nominally 10 years of efficient service in the RAF reserves. The number 10 was weighted upward by two for WW2 service. Stan was awarded the "Air Efficiency Medal" on 18 September 1947, Fig. 4.6, shown beside.

The letter telling him this and his entitlement to wear the ribbon, not the medal, had an asterisk added to say that as they were out of ribbon, and it would come later. Another letter in the family archive accompanying the arrival of the actual medal took another 3 years to arrive. It seemed the efficient men had left the Air Force.

How did Britain win the war!

Cpl Henry Sydenham, who had enlisted directly into the R.A.F. (not the A.A.F.) at the same time, was granted a Good Conduct Badge. Fig. 4.7, on 26/7/42 being listed as 1st and under A.D.R. and 'A', whatever all that means.

Fig. 4.7 RAF, *For Long Service and Good Conduct*, medal, WW2.

Those two experiences give some detail of life as a member of No 1 Balloon Centre; but there is so much more to appreciate. Finding RAF Kidbrooke Barrage Balloon men or women who are still with us and who served on the site, has not been at all fruitful. Their time with us has passed!

It looked promising when Nancy Pines, BBRC Member No 445, advised the BBRClub that she trained on Site 20, RAF Kidbrooke as a Balloon Operator in 1942, Fig. 4.8.

She wrote, in a follow up letter of 13 May 2013 to the author, 'I was only there for 3 months training, didn't go to the buildings and stayed in a hut.'

Fig. 4.8 Nancy's team; twelve WAAFs and two RAF men.

Note: This site number needs a prefix 1/, 2/, or 3/ for the Squadrons 901, 902 and 903 - to identify it and its precise location. Nancy said that their site is now under a post war road. A detailed story of herself and her group, mainly on the time they were stationed on Site 20 in Wandsworth, is given in the Christmas 2012 BBRClub Newsletter.

Next, contact was made with Percy Featherstone, Fig. 4.9, because of some chasing to locate someone with direct knowledge of uses of the various buildings then existing. He provided details of the site as it was some 4-5 years after the war years. Percy was not in 'Balloons' but in 'Motor Transport', on the RAF Kidbrooke site.

Fig. 4.9 Percy Featherstone with the bus he drove in the 1950s at RAF Kidbrooke.

'I (that is, Percy Featherstone) was there in the Motor Transport MT motor pool providing transport for everything and everywhere in and around London. There were many civilian drivers and fitters working there, but few RAF personnel.

Four of us were posted from RAF Weeton, used for technical training, Fig. 4.10, to there in Dec 1951. We were fitters, but two were eventually posted on to RAF Yeadon, now the Leeds/Bradford airport. The two of us that were left re-mustered to being a driver/mech, so spent the rest of our time mainly driving coaches.

Fig. 4.10. RAF Weedon was sited here until 1965.

Others in our billet were despatch riders and personnel who worked at the recruiting offices up in London; they just slept here and used it as their home. The camp was split in two. The bulk of activity was on one side of Kidbrooke Park Road with the Officers Mess on the other. Next to the railway boundary we built a few concrete garages for officer's cars.'

The BBRClub website, in *Formation of a Balloon Depot at Kidbrooke,* explains the site's birth in 1937 for use by the end of June 1937. [Destiny? That was right in the month I was born!] An early plan of the buildings needed was found, Fig, 4.11. It appears that quite major differences resulted as the authorities worked out what was needed and how to fit that into an already used site with some farmland added.

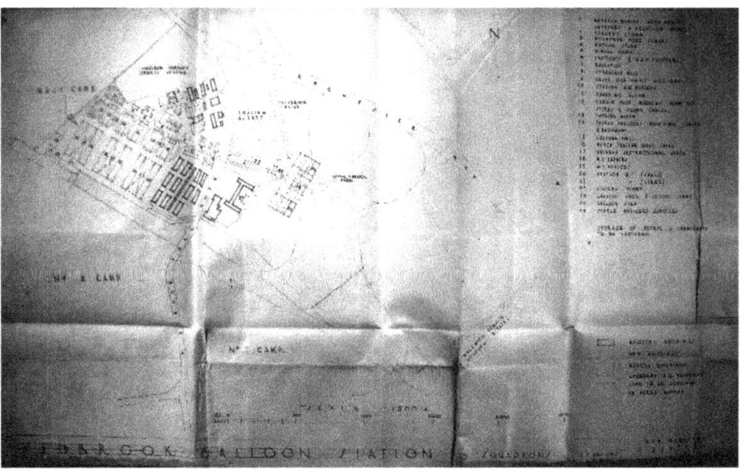

Fig. 4.11 First plan of No I Balloon Depot.

When Balloon Command ceased to exist after February 1945, use of the Kidbrooke base would have been changed to possibly provide more storage of surplus military equipment, ongoing transport needs and new uses.

Another statement of F L Jones, ex RAF says:

'I was posted to No 4 MT Company RAF Kidbrooke in 1946. The camp was a vast motor pool providing transport for service personnel transiting through London main line railway stations as well as providing transport for the Air Ministry, etc. All kinds of vehicles there; from staff cars to troop carriers. Over the railway bridge Kidbrooke Park Rd was No 1 MU, RAF publications. Flats are there now.'

A detailed usage plan of buildings in the balloon era has not been located for RAF Kidbrooke. Anyway, it would probably be complex due to the multi-purpose use of many old buildings.

The last contact I had with a person who had family at No 1 Centre was with Brenda Parsons, the daughter of Cpl Clarence Happs, known better as Bill. This is part of the sad side of the Centre's history. On 18 June 1944 Bill was killed when a V1 bomb made direct hit on his hut. That event, taken up later in this book - was the stimulus to research this story of No 1 Balloon Centre.

Brenda (also known as Hattie) sent details of his funeral and burial. Bill was given a fine funeral in his hometown. Hattie was just

2yrs old at that time, so has no memory of it. He was buried in the churchyard of St Thomas's Church, Sutton in Craven, Yorkshire. The headstone reads:

"343594 Corporal C W Happs Royal Air Force 18th June 1944"

Fig. 4.12 Bill Happs.

A fuller account of Bill on that fateful morning is given in Chapter 9.

The set-up details of each Squadron are given in their individual Operational Record Books ORB.

- 901 was formed on 16 May 1938 at the Kidbrooke site.
- 902 had its first mobilisation exercises at Kidbrooke on 22 January 1939. Its band was set up on 5 March 1939. Mobilisation of mainly AAF men into RAF embodiment was a gradual process at 25% of the target need, per month.
- 903 was formed on 16 May 1938 at the Kidbrooke site.

(Note: According to the BBRClub website article *"1940 Status - Equipment and Location of Balloon Squadrons* there were also Squadrons 952 Sheerness of November 1939, and 961 Dover of July 1940, attached to No 1 Balloon Centre. These included some 40 waterborne balloons. They are not covered here. That statement also does not include 903!)

RAF Squadrons could have a badge and motto by Royal assent. On 2 August 1939 the 902 ORB reports approval of its own badge:

'His Majesty the King has graciously approved the badge and motto for the Squadron. The badge is a Plant of Sunden. The motto is "Hostibus Obviam" which means "Against the Foe,'

Fig. 4.13 Was this Sundew insect-eating plant the paradigm of the crest of Squadron 902?

Images of that badge and what the 'Plant of Sunden' is for this, and of any 901 and 903 badges, have not been located or found on the Internet. It seems more likely that 'sunden' could be 'sundew', an insect eating plant with several long tentacles pointing up to the air with sticky ends just waiting for an insect to fly into them - very fitting for a balloon barrage image, Fig 4.13.

The next chapter investigates what was needed to set up a balloon centre.

Chapter 5 Make-up and Roles of a Balloon Centre

What were the space requirements to operate a Balloon Centre? It might seem to be a simple, straight forward, operation! Upon study of sites, for which we do have records, that assumption rapidly gets blown into the sky. The operation was major! Let us start with the balloons.

'The standard barrage balloon used throughout the war was designated the LZ (Low Zone). It was just over 62 feet (18.9m) long and 25 feet (7.6m) in diameter at its widest part and had a hydrogen capacity of 19,000 cubic feet (538 cubic metres).
The LZ balloon was flown from a mobile winch and was designed for a maximum flying altitude of 5000 feet (1524m). The winch speed limited the raising and hauling down speed to about 400 feet (122m) per minute, which meant that the balloons required 11 minutes to reach 5000 feet from their close-hauled altitude of 500 feet (152m).'
www.epibreren.com/ww2/raf/902_Balloon_squadron.html
John Penny, *A short history of No.11 Balloon Centre at Pucklechurch*, 1939 to 1945.

Balloons were made from rubberised cotton. They were coated with aluminium powder to dissipate heat from the sun and to lengthen the life of the rubber. For transport and storage, the balloon packed down into a relatively small package that was carried to a site on the back of the winch truck. The packed-up balloon could be managed on to, and off, the truck by 3-4 people without hoist lifting assistance.

Paul Francis, at www.airfieldinformationexchange.org, records:

'At the balloon depots several specialist buildings not seen before were provided, notably the distinctive balloon sheds and the unusually large motor transport facilities to house the fleet of winch vehicles. Provision was made too for training facilities and a balloon exercise area as well as buildings for fabric repairs and the storage of packed balloons. Accommodation was planned partly in light steel construction for the larger technical buildings and in type 'A' hutting for other technical and domestic buildings.'

Sending up balloons may seem, on the surface, to be a simple task to organise with its balloon winch, and gas cylinder trailer. However, full time operation of a squadron needs considerable overhead and ancillary support.

Balloons could be quite contrary if the winds took over as can be seen in a Pathe film of *1940 Balloon Goes Berserk - October 1940* with the explanation:

'Ground to air shots of a damaged barrage balloon twisting and turning on its leash high above London. People watching from street.' Pathe Film ID 1655.03.

Whilst not necessarily filmed at RAF Kidbrooke and having a now politically incorrect description of the WAAFs involved, a useful Pathe film shows some of the operations of balloon launching. Its description is:

'Various shots of Balloon Section of the WAAFs (Women's' Auxiliary Air Force) standing to attention, a female officer inspects them. The women are all dressed in boiler suits and are being trained in handling of barrage balloons, to help release more men for RAF operational duties. Narrator describes them as "sturdy types" who in peacetime would be "Gym mistresses, Policewomen or mothers-in-law". Various shots of women filling a barrage balloon with gas using a hose. Various shots of girls pulling ropes of balloon as it is launched into the air. No shots of the balloons actually in the air.' Pathe Film ID 1099.37.

Women took their part alongside the men.

Recruitment posters, Fig.5.1, usually did tell the truth about the glamour side but left out showing the privation they had to take carrying out their roles.

Fig. 5.1 WAAF recruitment poster.

No 3 Balloon Centre and Balloon Command HQ, at Stanmore, Essex, Fig. 5.2, becomes relevant here because of its site map providing a list of uses for buildings. Being purpose-built it gives an idea of the needs to be met at the mixed use, RAF Kidbrooke, Fig. 5.2.

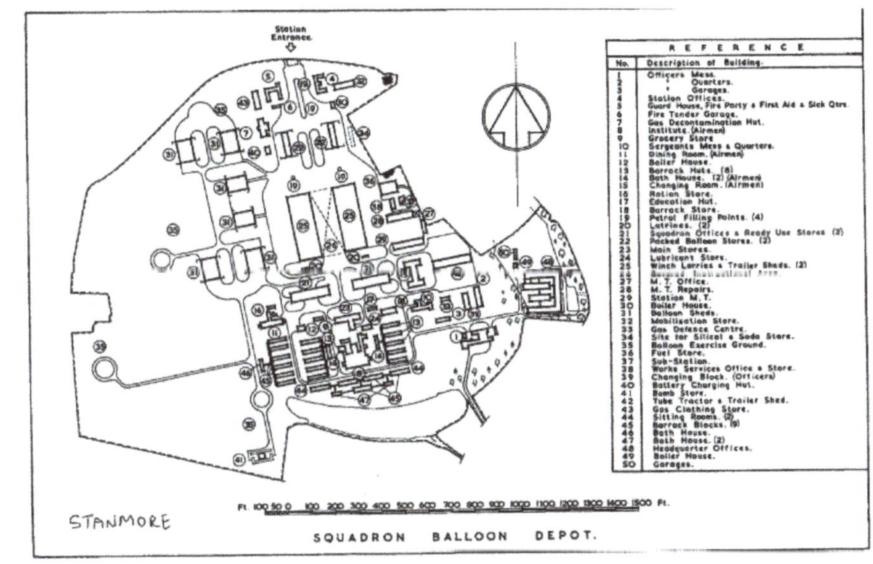

Fig 5.2 Stanmore, purpose designed, balloon centre.

R E F E R E N C E	
No.	**Description of Building.**
1	Officers Mess.
2	" Quarters.
3	" Garages.
4	Station Offices.
5	Guard House, Fire Party & First Aid & Sick Qtrs.
6	Fire Tender Garage.
7	Gas Decontamination Hut.
8	Institute.(Airmen)
9	Grocery Store
10	Sergeants Mess & Quarters.
11	Dining Room. (Airmen)
12	Boiler House.
13	Barrack Huts. (8)
14	Bath House. (2) (Airmen)
15	Changing Room. (Airmen)
16	Ration Store.
17	Education Hut.
18	Barrack Store.
19	Petrol Filling Points. (4)
20	Latrines. (2)
21	Squadron Offices & Ready Use Stores (2)
22	Packed Balloon Stores. (2)
23	Main Stores.
24	Lubricant Store.
25	Winch Lorries & Trailer Sheds. (2)
26	Covered Instructional Area.
27	M. T. Office.
28	M. T. Repairs.
29	Station M.T.
30	Boiler House.
31	Balloon Sheds.
32	Mobilisation Store.
33	Gas Defence Centre.
34	Site for Silical & Soda Store.
35	Balloon Exercise Ground.
36	Fuel Store.
37	Sub-Station.
38	Works Services Office & Store.
39	Changing Block. (Officers)
40	Battery Charging Hut.
41	Bomb Store.
42	Tube Tractor & Trailer Shed.
43	Gas Clothing Store.
44	Sitting Rooms. (2)
45	Barrack Blocks. (9)
46	Bath House.
47	Bath House. (2)
48	Headquarter Offices.
49	Boiler House.
50	Garages.

Fig. 5.3 Stanmore requirements.

Fig. 5.3 is the list given with the Stanmore plan. The good version of this plan is found in *The Royal Air Force Builds for War - a History of Design and Construction in the RAF, 1935-45*, MOD, 1997.

The list of functions provides insight into the numerous facets needed by a Balloon flight. They included:

- Winch trucks, Fig. 5.4, various models were used, for detailed use see *Taking a Look at Winches,* Dave Wintle, BBRClub web site.

Fig. 5.4 Fordson balloon winch truck.

- Hydrogen cylinder trailers, Fig. 5.5.

For No 1 Balloon Centre, they were filled using a dedicated hydrogen generator situated on the Kidbrooke site.

Fig. 5.5 Gas bottle trailer.

Ancillary support services in a Balloon Centre were:
- Ambulances.
- Sick bay.
- Crew and support transport trucks.
- Staff cars.
- Dispatch (also 'despatch') riders and their motorcycles.
- Meal preparation facility and delivery to sites.
- Accommodation for 24h operation duty staff on site.
- Store – general.
- Various kinds of garages.
- Balloon stores and repair sheds.
- Toilets and ablution blocks.
- Air raid shelters.
- Fire station and tenders.
- Motor vehicle maintenance garages.
- Maintenance staff offices.
- Gas stores.
- Boiler house.
- Coal storage for boiler.
- Silica and soda storage area - used to form hydrogen.
- Hydrogen gas generator plant.
- Officer's mess.
- Sergeant's mess.
- Other rank's messing.
- Fuel store, filling, and despatch pumps.
- Training rooms and sheds.
- Mobilisation shed.
- Headquarters staff offices with logistics and stores control, pay office, CO office, records of servicemen and women, post

office with censorship.

- Short term jail rooms.
- Parade ground and balloon exercise places.
- etc.........

There was also support for photography. Individual's Service Records have entries of "Annual filming done".

Tantalisingly, the 902 ORB Appendix contains eleven, 30-person, group photos. The groups are taken at five different background locations. They are numbered as 2,2,3,3,4,4,5,5,6,6,7. That number does not fit with a balloon squad or a Flight. That some high-ranking officers are included suggests they may be the HQs staff for the five Squadrons and the Centre. They are all men. The photos were designated by a directive dated 31 May 1940, suggesting they were taken prior to that time.

These pictures are of dubious quality for recognising people in the digitised download copies; the originals appear to be present in the actual OR Book. There are no captions or reference to these ORB images. Maybe the airmen are named on the reverse of the originals!

The Stanmore, a contemporaneous purpose built, site had to support the Balloon Command Head Quarters and two Squadrons 906 and 907. From the layout seen above it is possible to estimate 140 acres (60 hectares) of land was needed.

The space requirements of No 1 Balloon Centre at Kidbrooke, with its 3 squadrons, and its own Command HQ and 3 Squadron HQs, plus some flight HQs would have been much the same size requirement as Stanmore. From this it seems reasonable to suggest No 1 Balloon Centre Kidbrooke, with its three Squadrons 901, 902 and 903, would have needed a site footprint of at least 100 acres (40 hectares).

From a 1951 plan, the whole balloon base called RAF Kidbrooke, had an estimated total area of 250 acres (101 hectares). This can only be a guestimate for the base used many buildings around all sides of the railway line and the Kidbrooke Park Road. Clearly, balloon work was a significant part of the whole base.

Unlike the 'green fields' site of Stanmore, before WW2 RAF Kidbrooke had been long established as No1 Maintenance Unit MU. Thus, a large amount of adaption, extension and upgrade was needed to fit in Balloon Centre No 1.

Each Squadron, such as 901, 'sailed' 4-5 Flights. Each Flight had around 9 balloons. At the Kidbrooke site there were the three 901, 902 and 903 Squadrons. Thus, No1 Balloon Centre had some 150 winch units, including some downed ones and other attending vehicles, such as large Crossley tow trucks.

For each of the three squadrons, all balloon Flights had 12 men (later 14 women) to operate each of its balloons meaning there were 1620 persons operating the balloons. They also needed extra persons for round the clock operation plus the host of ancillary requirements.

All up, it is reasonable to estimate No1 Balloon Centre needed all manner of facilities, Fig. 5.6, to support over 2500 persons and their stores, etc. Not a small operation! This tally matches the ORB reports of Squadron strengths.

Fig. 5.6 Grub Up time. Balloon squads were often fed from their HQ with hayboxes.

These people did not all work on the Kidbrooke site for many lived around the SE of London, on or near, to their operated balloons. They were accommodated in all kinds of places - warehouses, billets in private dwellings, tents, huts or portable vans on site, air raid shelters, and more.

A balloon team carried out packing the trucks with stores, loading the balloon in its bag onto the winch truck, maybe coupling up the cylinder trailer, and driving to the site. Often this was needed to be carried out in black-out conditions at night, then unfolding, inflating, launching the balloon to the required height, pulling it down at desired speed, topping up the balloon gas, deflating it - and then there was often a need for spot repair of simple failures.

Doing these functions in the rain, sleet, hail, and snow; under windy conditions on an open common site in the dark with scant illumination, was the challenge to teams and to their health. On arrival at a site, they often needed to lay telephone cables to provide contact with their Flight and Squadron HQ.

Two members of balloon flight, one or two being a Corporal, were certified Balloon Operator AC 1 and AC 2 crew who also had to drive the different types of trucks; possibly go fetch meals from the nearest Balloon Centre or local cook house; and endure tedious guard duty periods at their flying site.

Sites, and the base at Kidbrooke, were often under enemy aircraft attack. That aspect entailed working in the worst of winter weather, too often changing balloon flight parameters to cope with wind, operational directive, faults and damage by aircraft and gun fire. It could be a tough and dangerous duty; certainly, it was not like flying a kite in the park, on a fine summer's day!

Some discussion of the life of a balloon team member is to be found in John Christopher's book *Balloons at War*, Tempus, Stroud 2004.

Len Bacon, BBRC website, records:

'A typical shift would start with a call to go on Guard Duty for two hours after which the operator would try and get 3 hours sleep. At 0700 hrs the Mess Orderly for the day would be required to ensure the whole Crew were awake, and ready to start the various duties required. After breakfast, having cleaned the billet, personal equipment, and rifle, they would proceed to the daily routine as follows:

- Operational Orders to fly, alter height, close-haul, or bed the balloons. (These numerous movements are recorded in the ORBs.)
- Switch on or off rip link.
- Daily inspections.
- Keeping balloon head to wind on the bed guy, mooring or interim close-haul.
- Maintenance of balloon: repairs and topping-up.
- Maintenance of winch: cleaning and brake tests. Winch tool maintenance and check.
- Maintenance of balloon bed: Blocks, tackle, wires, pyramid, cradle, sandbags, slips, rag bolts, "U" bolts, ringbolts, screw pickets, 90-foot circle and strops, tail guy snatch block, central anchorage snatch block.
- Maintenance of flying cable: oil and inspect throughout. (Done with heavy greasing by hand with a chunk of rag dipped in a grease pot!).
- Inspection and maintenance of armaments.
- Gas drill.
- Defence and weapons training.
- Lay-out of kits and bedding.

- Maintenance and cleaning of personal kit.
- Messing fatigues and on-site cooking.
- Inspection and maintenance of gas cylinders, trailer, topping up and inflation equipment.'

Accommodation at sites was initially tents, and rations were sometimes supplied via hayboxes, from central cook houses. Later hutments were provided, and food could be cooked on the premises.

They did get some personal time during the day and some, several days long, leave periods. They could then go to their designated billet. Cpl Henry Sydenham Service Record stated his billet address was at the Rectory in Patching, near Worthing on the South Coast. Why there is a mystery? It was miles from public transport and in the direct flight path of enemy planes and later V1 bombs. One wonders if he ever used it! It does not necessarily mean he stayed there; it might have just been a mailing address.

Not all training at RAF Kidbrooke of operators took place on the base. Nancy Pines, one of the few veterans in the BBRClub, see Chapter 4, has published her training experience at her site, Fig. 4.8.

Balloon Centre No1 at Kidbrooke had a top-notch dance band called the *Skyrockets*, Fig. 5.7. The conductor (Paul Fenoulhet) looks so alike to Cpl Henry Sydenham who was a magnificent by ear pianist but could not read music scores!

Fig. 5.7 The RAF *Skyrockets* band.

The band appeared over their time, before and after the war years, with such celebrated stars as Carmen Miranda, Frank Sinatra, and Danny Kaye.

The text with this picture of the band, Fig. 5.7, on the website tells how they came to be in RAF Kidbrooke:

'Although the original ten players (Added: from different top bands) of the Skyrockets volunteered for the R.A.F. with the

object of being selected for the Military Band of No. 1 Balloon Centre, Kidbrooke, they were not enrolled as musicians.

Fig. 5.8 Chick Smith started the Skyrockets.

They had to qualify as tradesmen in Balloon Command before being able to settle down to their music, which came second to active service. The whole idea started when trumpeter Chick Smith, Fig. 5.8, who at the time was with Syd Phillips at the Le Fuive Club in the West End of London. He heard about the proposed military band at Kidbrooke. He visited Kidbrooke where he discovered that first class musicians, who volunteered for the R.A.F. at once, could be directed to Kidbrooke as soon as they had completed their initial training. Chick spread the news among his friends and one by one, ten of the finest British dance band musicians went along to the R.A.F. Recruiting Depot at Eltham Hill to enlist.

Shortly before Christmas, 1940, Chick Smith, Paul Fenoulhet, Bill Apps, Alf Horton, Cliff Timms, lzzy Dunman, Jock Purvis and Pat Dodds were posted to Kidbrooke. Les Lambert and Jock Reid were erroneously sent elsewhere but arrived a few weeks later. All ten went into the Military Band of No. 1 Balloon Centre, conducted by Corporal George Beaumont, out of which, in February 1941, emerged a twelve-piece dance orchestra subsequently to be called the "Skyrockets," a title suggested by the Band Officer.

When Corporal Beaumont left Kidbrooke in June 1941, Paul Fenoulhet was elected his successor. While at Kidbrooke the Skyrockets did guards and fatigues and serviced the great silver balloons, which obstructed the approach of enemy aircraft. In December 1941, the Skyrockets went into the Army and Air Force radio show, "Ack-Ack, Beer-Beer" alternating with the Royal Artillery (Woolwich) Band every other week until February 1944. [Added from Wikipedia: Programming was developed for specific services – "Ack Ack Beer Beer" for the anti-aircraft and barrage balloon stations, "Garrison Theatre" for the Army, "Danger - Men at Work", "Sincerely Yours, Vera Lynn" and "Hi Gang" for the forces generally.]

Their first public appearance was in the "Jazz Jamboree" - significantly at the London Palladium in September 1941.

[Ed. Fig. 5.9 shows they played at a top London theatre.]

Fig. 5.9 The London Palladium was THE place to appear alongside other great bands of the time.

Their first Sunday concert took place at the Embassy Theatre, Bristol, in October 1941. But mostly they were called upon to entertain the troops in khaki and the two shades of blue. North, South, East, and West they travelled, rough and ready, perched on upturned boxes in bumping lorries but counting it a privilege to brighten the off-duty hours of soldiers, sailors, and airmen, sacrificing their own leisure to do so. All the "Skys" were demobilised within a month during the autumn of 1945. They decided to stick together in "Civvy Street" and work as a co-operative band with Paul Fenoulhet as musical director and Les Lambert as manager.'

By January 1940, that is before the Blitz started, Balloon personnel across the country and overseas had risen to 40,000 men. Women were then soon given the job as well.

To further illustrate life in a Balloon squadron we now move to look at the effort needed to move a squadron from the South of England to Scotland.

Chapter 6. Squadron 992 on the Move

The 23 minute long, You Tube movie, *Squadron 992 (1940),* is worth watching. It shows how a RAF balloon barrage team was taught the ropes and process of getting balloons into place at a given site.

It clearly was made as a propaganda film by the General Post Office GPO film unit; being released in June 1940. But as such, it does tell a good story. Here is given a precis of that film, some extra material being added where it helps paint the situation better. All pictures in this chapter are stills taken from that source.

Balloons were flown to protect urban and industrial areas from bombing. By early 1940 it was clear that Britain was not going well in its battle with the rapidly approaching German army over in Europe. The people needed assurances that they were being well protected against the heavy bombing that would surely begin, very soon once Dunkirk had fallen. The film was directed by Harry Watt and uses RAF officers and men as its main cast.

It opens with an RAF officer addressing recruits, Fig. 6.1, on the purpose of a barrage of balloons.

Fig. 6.1 The class begins.

Dive bombing was expected as it is the most accurate method of hitting specific targets.

Group members are introduced as being from all walks of life and that a person with any trade or profession can do this job.

They are told about the structure of a balloon and its parts and what kinds of tasks are required; they get instruction and learn skills in roping knots and weaknesses, wire and material rope repair and splicing, how to make up a harness, and how ropes are attached, by gluing and stitching, to the sides of the rubberised material.

The officer tells them that a balloon can be hard to control and needing constant care – "far worse than minding a baby" - as it must be kept fed with hydrogen top ups, set to varying operational heights and taken down as storms approach.

They then get, outside of course, how to handle a balloon on a winch especially in wild, dark, and cold weather. They then drive balloons in and out of the huge hangers using the winch trucks. Fig. 6.2.

Fig. 6.2 Taking balloons out of the huge shed.

You will note that they usually wear their heavy great-coats! They were told that their balloon may well be given a name.

In the Operational Record Books, balloons that become unable to work are called a 'casualty'. The daily causality list for a squadron is typically 10 or so!

They are ready to go to work. A worrying event spurs the leaders to move a squadron to defend the Forth Bridge and the ships resting in the Firth of the Forth. A German bomber flies over the area dropping bombs on local targets. The raid is soon foiled by the magnificent spitfires! The bomber is shot down. A German crew of the bomber are saved from the sea, Fig. 6.3, by local fishermen who treat them well.

Fig. 6.3 Crew members of the downed German bomber are rescued.

The threat is apparent! A squadron of balloons is needed there urgently. Within an hour the order is out for this to happen. Throughout, humour is injected to soften the stark situation.

The convoy is being built up to leave in around 24 hours. Under the SECRET stamp, Operational Orders are issued, to those who need to know. The orders detail the kit, right down to the number of socks and food packs that each man must be issued with from the quartermaster store. The convoy must also carry spares and maintenance men; Squadron 922 is no longer close to a mother base to operate from.

It is long and moves slowly, but relentlessly.

Planning for this needed rapid action, Fig. 6.4. An advance party from Balloon Command, led by officers, visits homes and farms looking for suitable places for each balloon site and for the many ancillary needs, such as stores, messing and accommodation.

Fig. 6.4 Officers plan the operation.

They talk with owners about the RAF needs and the compensation that will be paid. Sites must be available on the spot.

They look for suitable locations and satisfactory road access to sites, a water-supply for each and, hopefully, firm standing. Food supplies and billets for officer ranks and digs for other ranks, including suitable fields for erecting tents must be located. This is all going on whilst the convoy is being provisioned and men made available, some being called back from leave.

The convoy drives all night to get there is major, being half a mile long; 50 lorries including winch trucks, gas bottle trailers in tow, trucks for 'other ranks' travel. There are 6-wheel trucks for the kit of the men, and essential repair and replacement gear. A despatch rider leads the convoy, setting the speed so that the last vehicle can keep up at the allowable night-time speed limit. The convoy stops for brief moments for food and 'rest room' relief. Officers travel independently by road in their staff cars driven by a transport person.

The convoy arrives around dawn; the men being tired out. NCOs are in place to lead trucks to the then selected site locations that had been marked with a simple number on a stake. Most of the men had never been to Scotland, or some 'even north of the Thames'. The fields and hedgerows look idyllic in the early morning sun. Tents had been erected on sites by the advance party. Gradually winch trucks, carrying a folded balloon on its back tray, reach their site. Its crew of twelve get to work infolding the balloon on the chosen site, marked with a tarpaulin and weight bags, and start inflating the balloon ready to connect it to the winch cable.

Fig. 6.5 Gas bottle trailer needs lots of grunt to get it into position

Getting trucks into sites, Fig. 6.5, was sometimes a challenge as the bottle trailers were eight and a half ton in weight - and the fields were not always solid. But they all get into place with the crew pushing, or a farmer using a tractor to move the trucks. Local folk are supportive and somewhat bemused at this technology arriving in their fields.

After just 48 hours the Squadron has been assembled, transported hundreds of miles up to the Firth Bridge, set up its five balloons and let them up to the operational height specified in their Orders.

Another 'wall of steel' is in 24hr operation. Raids on the Firth will be harder to commit for the Germans.

Crews move into their accommodation to get some well-earned rest, hopefully not always in tents. Officers go to their billets.

The commentator states "Within 12 hours the barrage is on active service" - as sentinels on the sky. Another effective contribution to the defence of Britain.

Fig. 6.6 Another balloon site is operational.

The next Chapter investigates learning the job.

Chapter 7. Learning the Ropes.

In the early 1930s, air travel over long distances concentrated on using hydrogen filled airships. 'Heavier than air' transportation was in its commercial infancy.

The size of the largest airships required all-weather hangars that could support them in their final ridged shape; they had delicate metal frames that held the massive hydrogen bags in place. Cardington had the large airship hangars of Britain on that site. It had been nationalised in 1919 as the Royal Airship Works.

The name 'Cardington' is mostly synonymous with the R100 and the R101 airship development, that began its design and creation life here around 1925. After its disastrous crash, in October 1930, large airship development in Britain ceased. It clearly was too dangerous for carrying people. Cardington then became a storage facility, looking for a new use, Fig.7.1

No 1 shed was, after extension, 35 feet (11m) high and 812 feet (248m) long. No 2 had been previously in Norfolk as a RNAS site and was reconstructed at Cardington airfield in 1928.

Fig.7.1 Cardington Hangars, Bedfordshire.

The need for hydrogen filled balloons re-emerged in 1936, this time in the defence of Britain. Barrage balloons began to be manufactured there.

Cardington became the No 1 RAF Balloon Training Unit (No 1 BTU); there were other training places. It provided the necessary training in balloon storage, manufacture, repair, and handling. Balloon operators and drivers were trained there. Here, also, the hydrogen needed for balloons was manufactured in its Gas Factory, using the steam reforming process. This process combines water and methane to create pure hydrogen and carbon dioxide. It needed the contents of 30 full hydrogen bottles to fill a balloon, and it had to be taken to the war sites with the initial set-up and by special delivery thereon; balloons leaked from tears and holes made by ack-ack shell fragments. Sometimes it has been said the only transport seen on the roads were balloon bottle trailers!

Balloons were also made in other places. Fig.7.2 shows this happening at a Dunlop factory in Manchester.

Fig. 7.2 Dunlop factory in Manchester.

In 1939 the No 2 Balloon Training Unit at the Rolleston Camp, Salisbury was dissolved with part of its staff and functions being transferred to No 1 BTU.

No 1 BTU continued to provide training of Balloon Operators and gave courses in balloon rigging, fabric work, motor transport, and balloon handling. When sufficient Barrage Balloon assets were achieved, the need abated; it played a reduced role from 1943.

In May 1941 a significant number of women started to flow through Cardington as they began to replace men in the role of balloon operations.

'Cardington would always be an ugly, cheerless place with its rows upon rows of Nissan huts intersected by concrete paths and long concrete roads, but a big joy to me was the little N.C.O.'s room that I had to myself at the end of one of the huts. There was just space enough for a bed and a locker and an old ammunition box, which went under the bed and in which I could keep my personal belongings. In the corner about three feet from the bed was a small iron fireplace. Each evening I lit my fire, which glowed red hot and filled the little room with a most luxurious warmth. I was very reluctant to go out in the evenings. I wanted to stay in and stoke up my fire. If the temperature dropped below 80F (26C) degrees, I began to feel chilly!'

WAAF Heather Simpson.

https://www.rafcardington.org.uk/WAAFBalloonUnit

Possibly the only commemorative artefact that recognised that the WAAFs were at Cardington is the WAAF Training Centre Shield, Fig. 7.3, recorded on the above website. 'Cardington' is engraved on the bottom cartouche.

Fig.7.3 Shield to the WAAFs who trained at Cardington.

Many attempts have been made to visualise a Barrage Balloon in 3D modelling for memorials. One attempt, Fig. 7.4, although not that distinct, was on this magnificent memorial cake baked for the third birthday of the WAAF units arriving at Cardington. The web image given here is lacking detail. The original image may still be available at the Archive Services Times & Citizen collection BP 1362/4 – photography: John Day Ampthill Image.

Fig. 7.4   Celebration cake of the Cardington WAAFs,

The final entry, in the Operational Record Book of this Unit, stated the following statistics of its use since its birth in 1937:

'The above final entries mark the cessation of existence of this Unit. Since its inception in 1937 No 1 B. T. U has trained over 5,000 R.A.F Balloon Operators: with a few exceptions every Balloon Officer in the Command has attended a Balloon Handling Course at least once, and many Officers have returned for a second course. About 5,000 W.A.A.F Balloon Operators and W.A.A.F Officers have been trained in balloon handling. The M.T school has trained as far as can be ascertained, about 8,000 Balloon Operator Drivers and nearly 4,000 W.A.A.F Drivers MT.'

www.Rafcardington.org.uk/No1Balloon Training.

During the war 23 Balloon Squadrons were based here. At the start of this at Cardington, only men were used to 'man' the balloons. Their training was covered in less detail in Chapter 6.

Whilst the films discussed here are about Cardington, training also was undertaken at major Balloon Centres such as No 1 Balloon Centre, Kidbrooke and No 4 Balloon Centre at Chigwell.

Fig. 7.5 '*A Balloon Site, Coventry, 1942.*'

WAAF topics were good publicity and there were, later in the war, more of them working the balloons, than did men. They also got more attention in the propaganda movies and posters.

Dame Laura Knight was commissioned by the War Artists' Advisory Committee (WAAC) to paint a study on barrage balloons. Fig. 7.5 was the outstanding result – with four WAAFs and an RAF man, under the leadership of the WAAF sergeant.

This, oil on canvas painting was undertaken in 1942. Coventry had been the target of a German bombing raid in November 1940, when over 10,000 incendiary bombs were dropped on the city. Apparently, the painting was intended to be a part of the recruitment drive for female operators, but it arrived too late to be used for that.

Fig. 7.6 Her composition started with preparatory studies.

Using a body of expressive drawings, she explored the barrage balloon picture she was to finally paint. Fig. 7.6 is one of them. Pathe films showed WAAFs in various steps of their training:

‘*WAAF Barrage Balloon School.* Visions of members of the WAAF (Women's Auxiliary Air Force) being trained in the operation of barrage balloons, Fig. 7.7. We see them in a classroom looking over the motorised pulley. We see them outside with a balloon, being shown how to tether it, roll it up, how to operate the gas tanks. We see the young women working with ropes and looking at a model balloon.’

Fig. 7.7 Using a scale model to ‘Learn the Ropes’

Fig. 7.8 Balloon WAAFs on parade at Cardington.

'Women from a variety of professions sign up to join a volunteer group looking after barrage balloons, Fig. 7.8. Working through all weathers and all hours, after an eleven-week initial training period, they are committed to their work as well as their leisure-time activities.'

Finally, on training, there is the film archived as IWM Record 1022; *Balloon Site 568, 1942.*

The film makers see the Strand Film in this way:

'A Tribute to Courage, Loyalty, and Commitment

Don't expect great art from a propaganda film like BALLOON SITE 568. It depicts the eleven-week initial training session of a group of women from diverse social backgrounds - a secretary, a homemaker, a factory worker - who decide to contribute to the war effort by maintaining barrage balloons. The training is quite arduous, involving listening to a lot of lectures accompanied by practical activities, followed by solemn lessons in the potential risks involved; all of them will be in the front line when the bombing of Britain resumes.

The women respond to their task with enthusiasm, realizing, no doubt, the significance of their roles. They receive willing help from local people, offering them food, baths, and shelter wherever possible, and learn to enjoy small occasions like receiving their often-infrequent rations.

We also see them enjoying their few moments of leisure-time in the dancehall, coupling somewhat self-consciously with male officers yet telling them of their inability to forge long-lasting amatory relationships. War work assumes far more significance in their lives.

As a sociological piece, BALLOON SITE 568 offers a fascinating insight into the way in which war work dominated ordinary people's lives, to such an extent that they never quite knew what was going to happen, even the very next day.'

The following is a precis of that film, with some extra information spliced in. The figures are screens shots from it.

It opens, Fig. 7.9, with a training group marching out of one of the Cardington hangers. They are dressed in worst weather gear - gum boots, sailor style oilskin hat and a long oilskin coat!

Fig. 7.9 Ready for action. WAAFs on Parade.

The film then moves on to show some typical women at their civies jobs; and who are soon to join up as WAAF Balloon Operators. Those shown as examples are a shop assistant as in Fig. 7.10, house maid, office secretary, barmaid, and later a stenographer.

Fig. 7.10 One person is a shop assistant.

At their joining interview they are told of the hard work and living conditions and of being on-call all the time. Their service will start with 11 weeks at a training centre, in the country.

They are issued with uniforms and necessities and are trucked out to live in wooden huts there at Cardington. On arrival they get a day to settle in before the rigorous training regime kicks in. The new recruits must have been overawed by the gigantic nature of the sheds the balloons being taken out for training; they had had never experienced this in their previous jobs, Fig. 7.11.

Fig. 7.11 Their first impressions of 'bigness'.

Classes start with an explanation of the application of the balloon barrage concept. They soon get into learning the details of the balloon they will, in a few weeks' time, be operating on a war site.

They use training items and the equipment needed to repair a balloon on site, Fig. 7.12. They learn many, new to them, rope

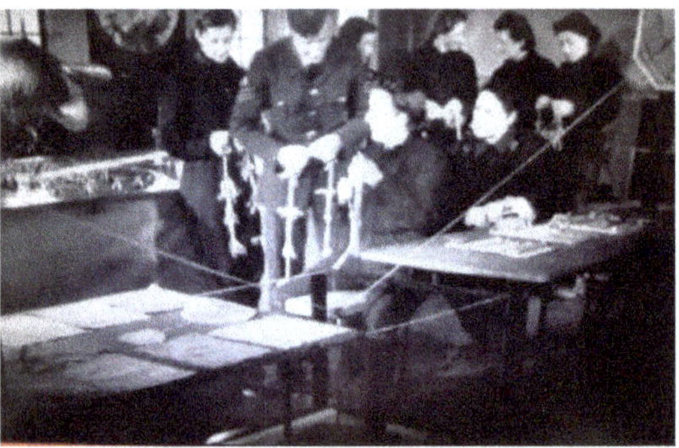

knots and how to splice together the multistrand wire rope that holds the balloon from the winch drum.

Fig. 7.12 Knot class time.

They are shown a winch and get a sample drive. Learning to repair winches and drive the winch trucks are skills they were most unlikely to have learned in their past day jobs. Weather conditions, at their worse, are discussed. The training is, however, not all slog. They enjoyed a social life at a dance with service men.

After the 11 weeks have passed, they are deemed ready to take an active part in the defence of their Kingdom.

Fig 7.13. Ready to tackle the hard weather.

Teams are sent to industrial locations outside of London. These young women, Fig. 7.13, are standing in the snow! They are ready to go to their designated war site, that could be anywhere in Britain. They are soon to be independent and responsible for their balloon, but the place and living conditions must be accepted; they need to be flexible. The team filmed get an industrialised location, war site 568, Fig. 7.14, with a cooling tower of a power station to protect. At least, this location has plenty of private houses with people who will look after then well in their off-duty hours.

Fig. 7.14 War site 568.

They have a small cookhouse nearby. An officer visits, Fig. 7.15, to inspect their food situation.

My father, who experienced similar situations in his Balloon Operator days in No 1 Centre, humorously summed those visits up with this comment: 'The officer would take a sip of the stew. Making a face of sheer abhorrence she would say quite forcibly; "Why! It is Delicious" '.

Fig. 7.15 Food inspection time.

Dry rations for a few days are supplied as part of the kit they take. In middle of well-earned relaxation their sergeant gives them their Orders to fly the balloon at 2000 feet. A corporal leads the team, with a much-used metal megaphone, Fig. 7.16, bellowing out Orders to 'let up' the balloon to the designated height.

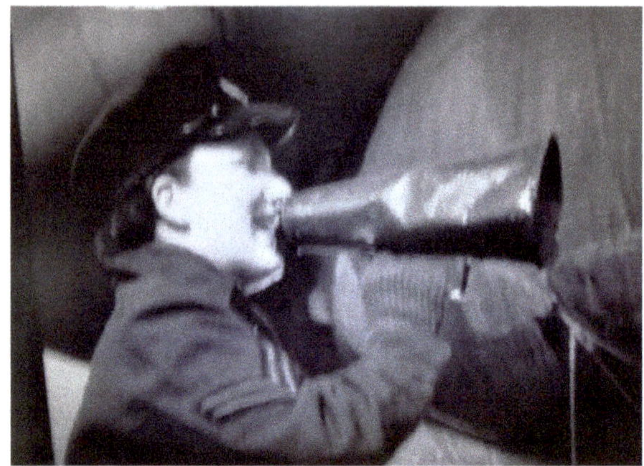

Fig. 7.16 "Let Up", shouts the corporal in charge.

War Site 568 is in service with the team waiting for another Order, or to cope with some yet unforeseen event that they can overcome. If too badly damaged, the balloon is sent to its base Centre for major repair, or replacement.

Chapter 8. From Operations Record Books

The ORBs

Each day, RAF Squadrons and Centres had to record the main events of the day in their Operations Record Book, ORB. Fig. 8.1 is the top part of the opening page for the 901 Squadron ORB. Page numbers are not used; the only index is by chronological date.

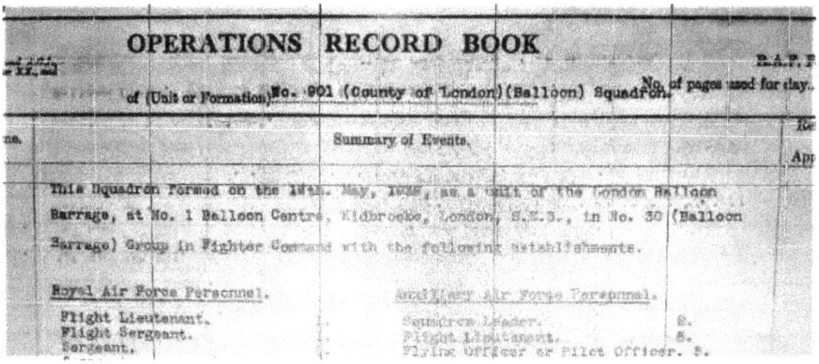
OPERATIONS RECORD BOOK

of (Unit or Formation) No. 901 (County of London)(Balloon) Squadron. No. of pages used for day.

Summary of Events.

This Squadron formed on the 14th. May, [illegible], as a unit of the London Balloon Barrage, at No. 1 Balloon Centre, Kidbrooke, London, S.E.3., in No. 30 (Balloon Barrage) Group in Fighter Command with the following establishments.

Royal Air Force Personnel.		Auxiliary Air Force Personnel.	
Flight Lieutenant.	1.	Squadron Leader.	2.
Flight Sergeant.	1.	Flight Lieutenant.	5.
Sergeant.	4.	Flying Officer or Pilot Officer.	8.

Fig. 8.1 Opening page of ORB for Squadron 901.

These records are held at the National Archives, Kew, UK and are available, for free, for signed-in members. Each ORB record is around 2GB in file size. They are only available as a download and come zipped up. They breakdown into pdfs for each of the above records. They are digital images taken of microfiche frames; the image quality is poor, but just about readable. They are not searchable for words, or dates, and so need to be inspected, page by page!

<u>RAF inventory number</u>	<u>Topic</u>
AIR/27/2219	901: full ORB Day entries
AIR/27/2220	901: and 910: joint Appendix
AIR/27/2221	902: full ORB Day entries
AIR/27/2222	902: Photos and Op Orders
AIR/27 2223	903: full ORB Day entries
AIR/27/2224	903: Appendix with Op Orders only

Those researched for this study are given above. There is probably an ORB for the No 1 Balloon Centre as a whole, but it was not obtained for this investigation.

An ORB covers many aspects of life in its Squadron. It contains, not only information about balloon operations and reports, but includes all manner of brief accounts of a range of aspects of Squadron life. There is, however, rarely a report of any length in the ORBs; other RAF records exist, for instance deaths during on, and off, duty.

Strength Reports

At monthly intervals the personnel strength (number of people) of each squadron had to be recorded. Fig. 8.2 is an example. They usually record nominally 700 persons for each squadron. Strength records started in 1938 with RAF and AAF men. From around February 1942 AAF men began to be replaced with WAAFs. These records end, when the Balloon Command ceased to exist, in February 1945.

Royal Air Force Personnel.		Auxiliary Air Force Personnel.	
Flight Lieutenant.	1.	Squadron Leaders.	2.
Other Officers.	1.	Flight Lieutenants.	5.
Flight Sergeants.	1.	Flight Lieutenant or Flying Officer.	3.
Sergeants.	5.	Flying Officer or Pilot Officers.	5.
Corporals.	10.	Warrant Officers.	2.
Aircraftmen.	50.	Flight Sergeants.	5.
		Sergeants.	10.
		Corporals.	48.
		Aircraftmen.	493.

Fig. 8.2 Strength Reports were entered at the end of the month.

The actual number fluctuated. On 31 July 1943, Squadron 902 had a personnel strength at 788 total, WAAFs being 245 of them.
In August 1943 it was 780 total, including 342 WAAFs.
In October 1943 there were 810 total including 447 WAAFs.
At the end of that year the total was 749 including 306 WAAFs.

There was a reorganisation of the RAF Barrage in August 1944; 80 WAAFs went from 902 to 961 Squadron, 32 to redundancy and 45 went to Cardington. Because of the V1 flying bombs, many balloon squadrons moved to the South Coast to stem the inward flow of V1s.

Personnel aspects

Postings and promotions of officers, and sometimes NCOs are given, Fig. 8.3. Rarely are names given of Other-Ranks persons.

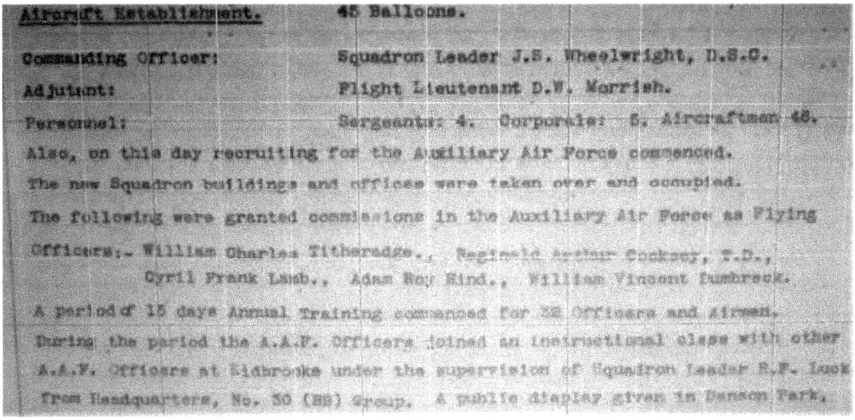

Fig. 8.3 Personnel records are frequent in ORBs.

Balloon settings and balloon casualties

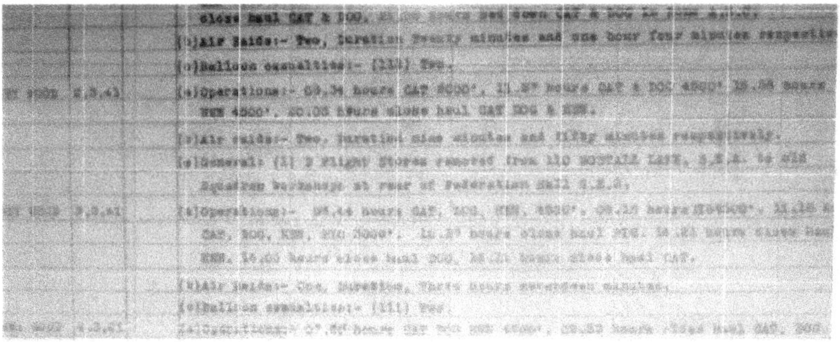

Fig. 8.4. Balloon orders fulfilled.

Fig. 8.4 is such a record. A non-functional balloon was deemed to be a 'casualty'. A dominant cause was natural lightning. When bad storms were predicted, balloons were pulled down and close hauled. They were not needed then as enemy attacks would also be lessened.

Another reason for casualties was that the gas diffused through the balloon skin; regular gas top-ups were needed. Another major problem were the sharp shards of steel that fell onto balloons when the locally fired ack-ack shells exploded in the area. Also, ropes would have unavoidable pulls that tore the connecting fairings.

Air Raid alarms

Numerous pages record air raid alarms, enemy bombers flying over the squadron area, bombings, and strafing, Fig. 8.5. These kinds of records greatly intensified during the Blitz (starting in September 1940), and the Baby-Blitz (late January 1944).

		P/O E.M. Niblock posted to 924/30 Squadron.
17.8.44		Barrage at C/Haul (25 balloons operational.)
	06.16	Alert sirens due to fly bomb plots.
	07.04	Raiders passed sirens.
	07.47	Alert sirens due to fly bomb plots.
	07.56	Raiders passed sirens.
	08.17	Alert sirens due to fly bomb plots.
	08.46	Raiders passed sirens.
	09.17	Alert sirens due to fly bomb plots.
	09.26	Site Q/48 inflated
	09.56	Raiders passed sirens.
	13.25	Alert sirens due to fly bomb plots.
	13.43	Raiders passed sirens.
	14.32	Alert sirens due to fly bomb plots.

Fig. 8.5 Frightening times at Kidbrooke in 1944.

To give an idea of the intensity of German bombing raids they used over 400 planes plus fighter protection each time, the main blitz lasting like that until May 1941!

Bombing all seemed to start up again with the surprise Baby-Blitz from January to May 1944 whereby Hitler ordered a last ditch, massive attack of over 400 planes, that then being most of the German bomber force left in service.

The records clearly indicate the periods when V1, and later V2, rockets came. A sample from a page is shown in Fig. 8.5. The ORB pages are just so full of warnings and seeing, or more likely just hearing the characteristic pulsating sound of their engine, as the V1s passed overhead at some 15-minute intervals.

This Balloon Centre was right in the path of V1s flying to inner London. These began to be recorded in early June 1944. At first, they were called "Rocket", "Fly bomb", "Jet", "Diver", "P' plane, "Raider" or "Plane". They became the V1 when it was found to be Hitler's name for them as "Vengeance 1". The Squadron 902 ORB ends on 31 August 1944.

V1s were still being recorded as 'Raider passed sirens". If you heard them pass over, you were OK. Although you knew you were safe from that one you heard - you would not hear the one that fell on you! When the engine ceased you could not be sure of where it would fall and explode with horrific damage resulting.

At the start of this horrific period, it was still felt that these V1s rarely fell in the Kidbrooke area, so Centre sporting events continued for a while! However, when the intensity of flyovers increased it did not take long for the V1s to fall on Balloon personnel. That is taken up in the next Chapter 9.

Training

Many instances record people going for training. When the AAF men were embodied into active service they needed more training. Due to large numbers to be brought up to full capability, they were sent 25% of the whole at a time. This began in June 1939.

The influx of WAAFs to replace men also increased the need for training in the 1942 and 1943 period. Three-weekly courses of nominally 30 operators were also organised at the No 1 Centre. They had to work hard at that learning. It was usual for a few participants to fail. Special training was provided, on 30 May 1944, for WAAFs taking part in a forthcoming Birthday parade.

Small arms training took place in July 1940. On 28 March 1941, Senior NCOs completed a series of full evening courses at the Scots Guards HQ, Elder Road, Dulwich, SE. Subjects given covered Mills hand grenade throwing, street fighting, concentrated firing, and visual training. They were repeated later. Site defence was not neglected. Guard duty was a required task.

Invasion of Britain by the Germans might come at any time; capability had to be kept in place.

Gas attack was also on the cards; men were sent to the Anti Gas School course at the Rolleston Camp in Larkhill.

Discipline
Several Court Martials took place, usually for those who did not return to base in time – called 'illegal absences' there. No personal names were recorded on the ORBs for these situations. With No 1 Balloon Centre personnel often living within easy reach of home, and homes and civilians being bombed so much, it would not be unusual for serving men to be pulled by their personal duty to serve their families more than the balloons.

Awards
Awards were made to serving personnel. Two cases recorded in the No 1 Balloon Centre ORBs are:

Temporary Corporal S Stephens, a Balloon Operator in D Flight of (possibly) Squadron 902, received a Commendation Certificate that was gazetted on 25 August 1941. This was for his duty carried out on war site 2/16:

'On many occasions this site has been very heavily bombed. This NCC showed courage and initiative of high order, encouraging the crew to extinguish incendiaries and fires often under very heavy bombing and AA fire.'

Another service award was to Sergeant Oliver Nather, who was granted the British Empire Medal (Military Division) on 27 January 1943. It is awarded for 'meritorious civil or military service worthy of recognition by the Crown'. It was below the level required for the award of the George Medal.

Public Works
The Centre was often involved with events of public nature.

There was an Annual Exhibition of toys made by Squadron 902 personnel. One recorded in an ORB was of that shown at Goldsmith's College on 14 December 1942. They donated 450 toys that were received by the Lord Mayor for London for distribution to children whose houses had been damaged by enemy action.

Salvage drives took place in war-torn Britain due to the Atlantic Blockade by Germany that was slowing up imports from overseas, Fig. 8.6.

Fig. 8.6 WW2 salvage drive poster.

Other recorded events in the ORBs were Salvage collections made by the Base. That of February 1943 produced more than 3434lbs (1.5 tons) of iron, steel, rubber, aluminium, silver paper, bottles, brass, paper, and other items. Often mentioned in salvage drives were empty toothpaste tubes and used razor blades.

No 1 Centre, as whole, had a salvage drive in January 1944. The total collected was around 84,513 lbs (some 42 tons). No doubt much of this had been hiding in the nook and crannies of the many, very old, buildings on the RAF Kidbrooke site. No doubt, the Kidbrooke base had accumulated much salvageable material from its long period since the 1910s of involvement with a wide range of heavy, medium, and light engineering fabrication and repair work.

The Skyrocket Band, mentioned earlier in this book, was another public contribution that made its name across the land.

Around May 1939 Squadron 902 took part in 18 performances in the *Royal Navy and Military Equipment Tournament* at the Olympia exhibition centre, Fig. 8.7.

Fig. 8.7 Olympia poster, 1939

The actual naming of this event is hard to resolve for in that month and year were held this event and the *Royal Tournament*. That suggests both names were used for this event. It would be interesting to know about that part the Barrage Balloonists played in this last tournament display before WW2 took place.

Sporting Activity
RAF and AAF men, and later WAAF women, working at the base or on its outlier war sites, had plenty of opportunity to partake in a wide range of sports. They still had to allow for war duties but there was much off-duty time for personnel, mostly then away from home.

Sports undertaken included swimming, net ball, cricket, tennis, soccer, snooker, table tennis, darts, and others.

These were played in the base where suitable, but they also took place in the Dulwich College grounds, and even once at Olympia. The Finals of the 902 Squadron Table Tennis Championship were played off in the HQ building in December 1943.

Overall management

The ORBs give names and postings of many senior staff, as of their promotion or departure. There was constant change in top-level management over the short life of Balloon Command.

Some major reorganisation took place as needs altered. On 28 October 1942, 902 joined with 903 to become Squadron 902/903.

The South London Barrage Control started up on 27 Nov 1941.

Socialising

Those who were available off-duty, attended a *Jack and the Beanstalk* pantomime held at No 1 Centre on 11 Jan 1944. No doubt that was a local amateur event, not that of a city theatre, Fig. 8.8. A girl usually played the part of the lead boy in these performances.

Fig 8.8 Vintage 1940 English poster.

Parties were held for personnel; often at Mrs Moat's home at 25 Champion Hill in Dulwich, a very old and distinguished area with large gardens and houses.

The custom on Christmas Day had the Officers and NCOs serving the other ranks and attending to their wants.

On Boxing Day 1943, a social evening was held at the Station HQ. It included wives, children, and friends, who were invited for an evening of entertainment.

On 31 Dec 1943 a social evening at Station HQ included a fancy-dress competition.

Operational Orders

A few of these classified statements are recorded in the smaller, second ORB for each Squadron. They provide considerable detail on operations.

Operational Order No 5 of 31 May 1940.

This one covered the actions to take should German paratroopers attack.

- The parachute troops are picked men who are quite unscrupulous.
- What to do about balloons 'keep the maximum number of balloons operating for as long as possible.'
- Fight the enemy and if you have to retire, destroy equipment that may aid the enemy.
- Flight Commanders to control actions.
- Conserve rifle fire and only on Crew Commander order.
- Use despatch riders, Fig. 8.9, and ration bicycles if telephone cables are cut.
- Keep HQ advised
- plus, more.

Fig. 8.9 Re-created WW2 RAF despatch rider with BSA.

- It also has retiring orders for where to muster.
- Ration bicycles to be ridden by crew members.
- 'Crews to be prepared to fight as they retire.'
- Plus, what to take and what to destroy, including the winches.

Order No 6 was given on 9 July 1940, No 7 on 23 July 1940, No 8 on 8 August 1940 and another on 29 July 1940.

Operational Order No 6 of 9 July 1940

This statement opens with:

'It may be necessary to provide a balloon barrage at short notice to protect a target that is bcing scvcrcly bombed by the enemy'

The intention is:

'To be prepared to move the Squadron or part of the Squadron as quickly as possible on receipt to orders. To inflate and fly balloons immediately in arrival at a new site.'

It then details how that is to be done using guidance from the already available *RAF Pocket Book*, Chapter II, Section 44. [1939, 320pp]. This little book is full of all manner of useful knowledge, Fig. 8.10.

Fig. 8.10 Pages from RAF Pocket Book, 1939.

The Operational Order then gives the top-level information needed by the Advance, Main and Rear Parties.

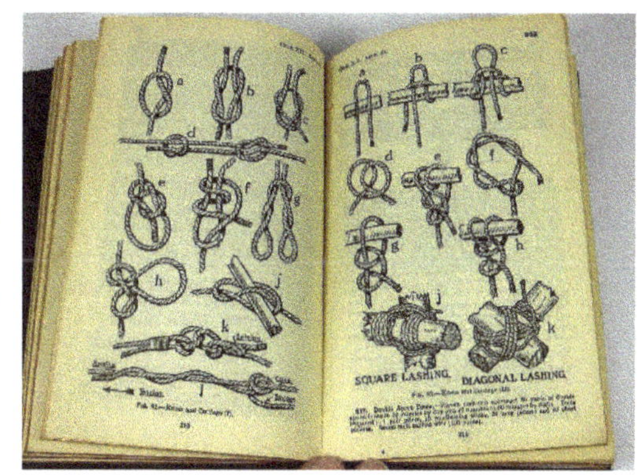

The Advance Party delivers the Commanding Officer and the Equipment and Balloon Warrant Officers. For reliable communications – there were no mobile phones or entirely reliable field wireless sets then - provided are 5 motorcycles with side cars, and presumably riders for each. A Senior staffer from each Flight is present and, finally, a Balloon Warrant Officer is in possession of six signal pads.

The Main Party sets up the Squadron HQ. A Sergeant Fitter, Transport Officer, Medical Officer, and Squadron Sick Quarters staff with an ambulance are included. Four telephone operators, two clerks and an Equipment Assistant complete the creation of the operational HQ.

Also in the main party are each Flight's crews. The Commander has his own despatch rider. There are to be a Flight Senior NCO, three telephone operators, and an Equipment Assistant, an NCO, a cook, and an assistant cook for the cookhouse. Then are listed crews that are moving, comprising the NCO and six airmen per balloon.

The Rear Party is small – just an Adjutant I/C, and per flight, an officer and SNCO, and any remaining personnel.

Then comes more details covering deflating balloons, packing them for transport along with requirements, as listed in their pocketbook. When packed, and the winches and bottle trailers ready, a report goes to the Flight HQ.

A long appendix list is given for the Cookhouse equipment.

The cooking centre hut is loaded onto a lorry, also carrying petrol and oil reserves. Next covered, is what to do with the yet unused food of that day, and the emergency iron rations. Hay boxes for an odd number of sites, are loaded onto a 15cwt truck (1.5 metric ton).

The Flight's 5cwt (0.5 metric ton) van carries the Flight Commander and senior NCOs along with the ammunition reserve. Another van carries the telephone operators, clerks, and Equipment Assistants. The transport office and the Fitter Sergeant go in another van.

The medical team with their equipment also used a small van to make the move.

Another long list is given of the gear that members of the main party are to be issued with.

Trailer winches are prepared and towed to the site, as is possible to use with the transport going. Extra towing vehicles will be used if available from No 1 Balloon Centre.

On arrival at a new site the first job is to inflate the balloons. Crews are to dig trenches, just deep enough to lay in, for sheltering during air raids. They also fill sandbags. As soon as the balloons are flying from their respective winches, the protection trenches are to be dug by all those available.

The main party uses kitbags for their personal gear and their No 4 gas mask, Fig. 8.11.

Fig. 8.11 No 4 gas mask kit.

All bedding at the base and other sites is returned to the Centre, to be sent with the rear party.

Blankets and palliasses, less straw, are rolled into bundles. Site fire appliances, ration bicycles and ladders are also to travel to the sites.

The accommodation being used by those moving away must be cancelled and keys returned to the property agents. The assistant to the Commanding Officer, (the Adjutant) seeks permission to close down, and take safe custody of the phone lines. He also advises No 1 Balloon Centre to board up the huts being vacated.

Transport for the operation is listed:

> 1 - 3-ton lorry per 3 sites = 15
> 1 - 3-ton lorry per Flight HQ = 5
> 2 - 3-ton lorries per Squadron HQ = 2
> 1 - 3-ton lorry for each trailer winch = 11

All up, 33 vehicles are needed to support the move.

Officers may use private cars to move; petrol coupons will be issued. They are permitted to carry additional kit in their cars.... a golf set perhaps!!!

And the final remark is that all crews are to be in constant practice at packing kits and winches.

There is no mention of where or how they all live on arrival at the site! It would be quite a task to find and authorise billets and other accommodations.

This Operational Order was signed off by the Wing Officer Commander – Commanding 902 Squadron, R.A.F, Kidbrooke.

To end this discussion on what can be learned from the content of the ORBs, here are a few snippets of interest.

- Trenches were dug at Kidbrooke at the declaration of war on 4 September 1939.

- A mobile squadron, known as "A" Mobile Balloon Squadron, was set up under Operational Order No 1 of 18 April 1940. Its HQ and waiting time were situated at Cardington. The text for this order has detail on the personnel involved, who are unusually named, and must have signed in for overseas service. Its crew of 18, coming from other sites when needed, had to be ready to move at 4 hours' notice!

- It also covers equipment in such detail as stating the number of men to share a pepper pot or a tin opener. Also detailed are their transport, rations, billeting and vacation of sites, fuel for 50 miles travel, route issues, medical examinations of the men and their inoculations and vaccinations, accounting of equipment and pay, the record sheets to go via despatch rider to Cardington. The order ends with a requirement to report their arrival at the next site.

Appendix A to ORB 903, (AIR 27/2224) 1940, has Operational Order No 1 for this, dated 18 April 1940:

- A Squadron, with all its trucks, equipment, and 40 men, left for Coventry on 15 October 1939. Detailed instructions for forming a mobile Balloon Squadron were given down to the last letter in Squadron 901 ORB Appendix.

- Squadron 902 at Kidbrooke reported being machine gunned by enemy aircraft on 27 July 1940.

Numerous entries are included on H.E. bombs and incendiaries being dropped on and near the base. On 27 December 1940 some 50 incendiaries were dropped.

The interactive 1940 'Blitz bomb site' tool at www.bombsite.org shows that some H.E. bombs fell on the base but attributes them all to being "near the Kidbrooke Park

Way". Many of the bomb hits noted in the ORBs of RAF Kidbrooke were, of course on external balloon war sites, not on the base itself. From that website tool it appears that the density of bomb hits on the base area was lower than around the local area. This is probably due to a lack of reporting from the base for security reasons.

- On many occasions windows were blown out, or in, numerous fires were extinguished, and many bombs exploded all too close to huts, buildings, and winches. Burning balloons often fell on sites.

- In late 1940, Squadron 902 had a nice 'rest house', Fig. 8.12, at the Manor House, Sundridge, near Seven Oaks. If the correct 'house' has been identified here it would not have then been the lap of luxury it is now!

Fig. 8.12 Sundridge Manor House, Seven Oaks.

- On 28 February 1941 a 'B' flight balloon took a steep dive due to winds. That caused its cable to fold under a car dragging it along, with its occupants, and then turning it over! *Balloons at War* by John Christopher, Tempus, 2004 has other tales to read.

- In February 1941 some records start to use codes with names 'Cat', 'Dog', 'Hen' and 'Pig' - presumably for the Flights of 902. No explanation for the names is given, but this method does not indicate to enemy intelligence how many flights are in use, as does numbering.

- A useful detailed war sites list for the balloons of Squadron 902, with street addresses to be used from 23 August 1941 to 1 August 1942, is given in the Squadron 902 ORB Appendix.

- On 20 January 1943 a seriously pursued attack was made on the Base by 6 to 10 Messerschmitt 109s, flying in formation. Fig. 8.13 is what the balloon crew probably saw as the formation was coming to fire upon them.

Fig. 8.13 Me 109 artwork depiction.

Numerous balloons were set on fire, falling over huts. Enemy aircraft cannon fire used tracers directed onto the grounded balloons. Nine balloons were destroyed on nine sites, half of which were operated by WAAFs.

- The ORBs also variously record the locations of Balloon Centre No 1 war site locations. They are sometimes listed for the initial set-up locations along with their 'cooking centres'. They are always not listed for easy extraction, an exception being where the Squadron 903 ORB has a single list, with street locations. Some locations given use the *Geographia*

London Street Directory grid reference system for the locations; but do not state which edition.

- For the Squadron 901's ORB the place of reporting changed from, 'Kidbrooke' to 'Abbey Wood' on 25 August 1939; where it says it is reporting for 901(B) Squadron. What this means is unclear. From that point on the record site locations are given for each of its Flights "A" to "E". The reporting location changes back to SHQ, RAF Station, North Camp, Kidbrooke, SE3 on 1 August 1942.

- On the Normandy invasion 'D-Day minus 1 day' the weather at Kidbrooke was recorded as "thunder and gale warnings" leading to high winds and rain by the evening. Much too bad to fly balloons; all 50 balloons of the reporting Squadron were close hauled. This was the kind of weather occurring just some 150 miles south, that almost caused cancellation of the D-Day landing on 6th June.

- As history now records, for D-Day 6 June 1944, the weather eased to be sunny, but still with high winds. Personnel at RAF Kidbrooke would have little inkling that on that day the Allies had stormed the Normandy beaches that morning.
  As the German bombers were rarely seen and were known to be beaten back, it was thought that the need for air defence would keep declining!

Chapter 9 tells what happened next; how the bombing threat upon London had not gone away, returning in a quite different way.

Chapter 10. RAF Kidbrooke after the War

Immediate post war use

Balloon Command ceased to exist in February 1945. RAF Kidbrooke carried on with the other purposes that existed there alongside No 1 Balloon Centre. A large amount of space was no longer needed. It was then mainly used for storage, maintenance, and training. RAF use of that land ended in 1965.

Since then it has taken 57 years to fully repurpose the site into future long term uses. It has had many uses, with the last significant area now under redevelopment as housing in 2025. The main occupants now are the Thomas Tallis school, Kidbrooke Park Allotments, James Removals, and in work, now the latest block (yet unnamed) of council houses on the SE corner of the site.

Percy Featherstone was stationed at the RAF Kidbrooke base from 1951-1953. In 2013 he marked up the plan he had already provided showing building uses then, Fig. 10.1.

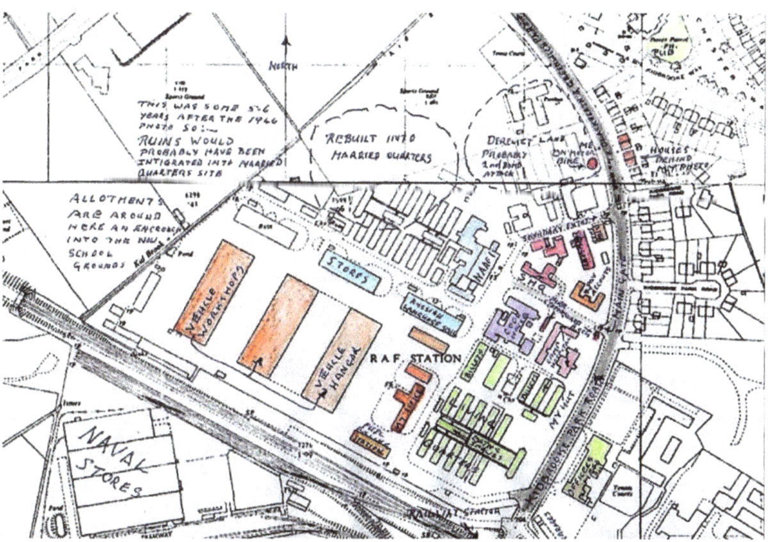

Fig.10.1 Uses of RAF Kidbrooke just after the war.

Percy was not in 'balloons'; he was with the Motor Transport MT pool that provided all manner of transport covering around London. His job was mainly driving coaches.

Civilian drivers, fitters, and a few RAF personnel, worked there. In Percy's billet were also despatch riders and recruiting staff who worked in London offices.

The officer's mess was separate from the main site, being situated opposite to the Thomas Tallis school's main gate, where now sits a supermarket. The very popular local pub, the *Dover Patrol*, was also there. It closed in 1994 and was demolished in 1995. From then on major redevelopments began that went on until the time of this revision in September 2025.

These historical details were supplied recently by Neil Rhind and David Wise:

'Its stores function was officially disbanded as a unit in its own right on 15 February 1947 but continued as a satellite managed from elsewhere, and it also housed some special units until early 1960's.'

Most of it has now been redeveloped, but some warehouses NE of the station are possibly in use as a store for Greenwich Maritime Museum, accessed from Nelson Mandela Road. A site NW of the station post-war, became the RAF Movements School. There was also from 1956-1961, training of RAF cargo-handling staff, and I remember the fuselage of a Hastings transport aircraft there from Jan 1959 to mid-1961, which was clearly visible from passing trains.

RAF No 4 Motor Transport Squadron was there until 1964. Its site was later partly used for Thomas Tallis school, and partly as a Post Office Vehicle Depot, then currently for warehousing. The main stores depot south of the railway had a railway siding accessed through a gate west of Kidbrooke station with its own shunter, diesel post-war but originally steam - two steam locos which are known to have worked here are still in existence, one in a museum in Leeds and one active on the Bluebell Railway. There was also a 2ft narrow-gauge railway within the site, including a bridge under Kidbrooke Park Road, subsequently used as a roadway when the site was redeveloped as the former Ferrier estate.

During WW-2 Kidbrooke also housed a variety of other units apart from its stores function. These included the No 2 Installation Unit, which was responsible for constructing and repairing the chain home radar station masts at many locations round the coast.

In WW-2, No 141 Gliding School for the Air Training Corps was on a different site nearby. It used a patch of ground, part of which survives as the Dursley Road ILEA Playing Fields. The gliding school there was in operation October 1942 to December 1945.'

RAF Intelligence training

From 1949–1953 the Joint Services School for Linguists was located at RAF Kidbrooke. Here Linguists were trained for covert work, their vigilance contributing greatly to National Security. The Thomas Tallis School, Secondary level, situated in the NE corner of the site, has a Blue Plaque, Fig. 10.2, on this use that was unveiled at the (first?) Thomas Tallis School in 2008.

Fig.10.2 JSSL plaque commemorating the Linguists' school.

During extensions to the school buildings the plaque was in the care of Martin Dean, Assistant Head Teacher (Operations) whilst waiting for its new location at the school after the extensions to buildings were opened in May 2012.

Replacing the Ferrier Estate with Kidbrooke Park
After 1965 several new uses were developed on that former base. Some of that land was then released to the Greater London Council for housing - the Ferrier housing estate. Built as social housing between 1968 and 1972; that ill-fated development became one of the largest and most deprived council housing developments in London.

Fig.10.3 A building is being replaced in 2012 at the Kidbrooke Village area.

After many difficulties the last tenants moved out in 2011 - whilst it was being demolished between 2009 and 2012, Fig.10.3. The 30-year Kidbrooke Vision scheme is replacing it with a massive multiphased housing and retail space known as Kidbrooke Village, a development of some 5000 homes. The 6-phase plan continues to be implemented, part of which is built on what was once Balloon base land. Fig. 10.4 is a sales view of Phase 5, that began in 2021 just south of the railway line. During the development of Kidbrooke Village, roads replacing the former Ferrier Estate were named to reflect the site history.

Fig. 10.4 Kidbrooke Village.

Kidbrooke Village got its first, brand new, pub in 2019. This community dining pub and watering hole, the *Depot*, is situated by the Kidbrooke railway station. Interior decoration commemorates the balloon era in one of its spaces, Fig. 10.5. On this wall is now mounted the Royal Aeronautical Society RAeS Heritage Award plaque – see Chapter 12.

Fig. 10.5 Balloons on the wall at the Depot pub.

Being within the Kidbrooke Village development it is publicised as a place to relax and work.

A new Kidbrooke railway station building, Fig. 10.6, was opened in 2012 to better cater for the 5000 homes of the Village.

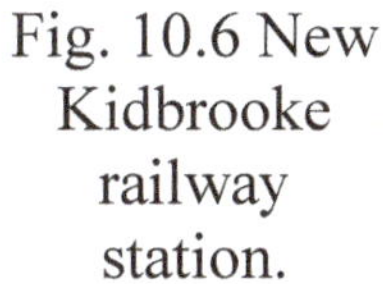

Fig. 10.6 New Kidbrooke railway station.

The remaining area, the school's first site, Fig. 10.7, in the SE corner of the former RAF site was approved in 2022 for building 117 new council houses as 19 one-bedroom, 60 two-bedroom, 22 three-bedroom, and 16 three-bedroom maisonettes. This project is much in work in late 2025.

Fig.10.7 Council houses will fill the remainder of the RAF Kidbrooke site.

That development completes the use of that part of the Base above the railway, filling it in between the allotments and school boundaries.

Kidbrooke Park Allotments

The 100 plot-holder gardening allotment, Fig. 10.8, on the land owned by the Royal Borough of Greenwich, is self-managed by the Kidbrooke Park Allotment Association KPAA.

Fig.10.8 Allotments resting by the removals building.

The KPAA was formed incorporating the existing Kidbrooke Park Allotment and Garden Society. The allotment association came into existence about 40 years ago.

Users said the land had been cleared and levelled, with manure being dug in to give the area good fertility. It has worked well, Fig. 10.9.

Fig. 10.9 Fertile allotments bearing fruit and flowers.

The Thomas Tallis school

The school was originally built in 1971, on the site now shown for development in Fig.10.7 above. Its former presence is still signposted, but for not much longer, Fig.10.10.

This, very large 11-18yrs age students' comprehensive school, Fig. 10.11, was completely rebuilt 40 years later as part of the 'Building Schools for the Future' programme using commercial participation with Equitix, being managed by G4S. This was achieved within the Private Finance Initiative PFI, a scheme whereby public services raise funds for capital projects from commercial organizations.

Fig.10.10 Reminder of the original site of the Thomas Tallis school

Fig.10.11 Thomas Tallis school.

In 1998 Thomas Tallis school was awarded 'Specialist Arts College' status. In 2005 it was awarded 'Leading Edge' status, being redesignated twice more. It was one of only 30 schools to achieve status of 'School of Creativity'.

The school's 50th anniversary celebrations were held in July 2022.

James Removals
The other major activity on the former RAF land is James Removals, who provide moving and storage services, Fig. 10.12. Their storage building is dominant, just by Old Post Office Lane, its walls being alongside some of the allotment sites.

Fig. 10.12 Removalist truck at the Post Office Lane shed.

Other uses.
Until recently, Recall Total Information Management was listed as being in the large storage shed.

Apparently, the area also hosted the Air Publications and Forms Store (APFS) at some stage, but that need would be almost obsolete today with forms and publications more likely to now be available on-line.

That ends an history of No 1 Balloon Centre, situated in RAF Kidbrooke during WW2. The last two chapters give here the Roll of Honour for No 1 Balloon Centre, followed by an account of the memorials to be seen in Britain for us to remember the Barrage Balloon operations that took place in WW2 Britain.

Chapter 11 Roll of Honour for No 1 Balloon Centre

The BBRClub website has a developing Roll of Honour, listing those known to have who lost their lives when serving in all the RAF Barrage Balloon Squadrons. The starting point for the list provided here was supplied by the Commonwealth War Graves Commission, via Mr Len Bacon. It may not be complete; other names have been realised in this study that were not on that list.

Those given here, extracted from the BBRClub list, are for RAF Kidbrooke Squadrons 901, 902 and 903 - i.e. No 1 Balloon Centre. Causes and places of death have been added, using entries about them in the three Squadron ORBs, and other sources.

The estimated number of those who served in Balloon Command is 33,000. Known deaths serving in Balloon Command are around 600. RAF Kidbrooke had 44 of these. With a death rate of 2 in a 100, this was certainly being in 'active service'. Not all deaths reported here took place on the Kidbrooke site itself, many of the lives were lost at their individual balloon war sites, around the London area covered by its 130 plus, balloons.

Dedication of the BBRClub National Memorial 2015

Name	Death Date	Rank	Service Number	Service kind	Age at DoD
Squadron 901 Operations Record Book					
WARD A. E. *Site 1/31 three bombs caused shelter dugout to fall in*	17/09/19 40	LAC	840231	RAF AUX	42
HOLLAND J.J *Reason for death, Heavy bombing of E Flight Site 1/20*	15/10/19 40	LAC	840423	RAF AUX	34
COLLEY J. W. *Killed off site in an air raid when on 7 days leave*	25/10/19 40	AC 1	640700	RAF	20
STUBBINGTON J. H. *Died from being knocked down by van in Abbey Wood*	24/12/19 40	AC 1	901395	RAF VR	37
GIBSON R.R.F. *No report in ORB on this death*	22/02/42	49	840190	RAF AUX	49
BROAD H.C. *Dive bombed hut killed 5 Site 1/22*	06/11/43	AC 2	1212586	RAF VR	-

GAUGHAN J. A.	06/11/43				
Dive bomber hit sleeping hut killed 5 persons. Site 1/22		LAC	1021188	RAF VR	33
GREASLEY C.	06/11/43				
Dive bomber hit sleeping hut killed 5 persons. Site 1/22		LAC	1026501	RAF VR	36
PHILLIPS I. S.	06/11/43				
Dive bomber hit sleeping hut killed 5 persons. Site 1/22		CPL	840330	RAF AUX	34
DODDS P.	18/06/44				
Direct V1 bomb hit on 901 HQ hut; 9 died; some later		CPL	863491	RAF AUX	32
ELDRIDGE E. R.	18/06/44				
Direct V1 bomb hit on 901 HQ hut; 9 died; some later		CPL	840009	RAF AUX	45
GENT E. S. B.	18/06/44				
Direct V1 bomb hit on 901 HQ hut; 9 died; some later		LAC	840084	RAF AUX	32
PEDRAZZINI J. C.	18/06/44				
Direct V1 bomb hit on 901 HQ hut; 9 died; some later		SGT	847696	RAF	37
SMALLEY R.	18/06/44				
Direct V1 bomb hit on 901 HQ hut; 9 died; some later		F/SGT	518861	RAF	-

HAPPS C. W. *Direct V1 bomb hit on 901 HQ hut; 9 died; some later.* *Transferred to 901 on 21 May 1944!*	18/06/44	CPL	343594	RAF	43
BARTLETT F.A.E. *Direct V1 bomb hit on 901 HQ hut; 9 died; some later*	23/06/44	AC 1	1666327	RAF VR	43
JACOBS J.P.H. *E Flight HQ demolished by V1 bomb.*	23/06/44	F/LT	87741	RAF VR	40
BURTON J. *(Aircraftwoman)* *E Flight HQ demolished by V1 bomb*	23/06/44	AC1	2025361	F.E.A. (?)	-
ROBERTSON W. C. *E Flight HQ demolished by V1 bomb.*	23/06/44	AC 1	1223474	RAF VR	34
WATERMAN F. T. *Died possibly from E Flight HQ V1 bomb incident.*	25/06/44	W/O	355294	RAF	40
MAYNARD G. C. *No mention in ORB*	20/09/44	AC 2	1864430	RAF VR	47

Squadron 902 Operations Record Book inc. 902/903 merger

BYERS L. *Not in BBRC list.* *Fell from truck in convoy. Died in hosp.*	22/05/39	LAC	528522	-	-

BELL R. *No mention in 902 Operational Record.*	28/06/40	CPL	512107	RAF	28
BOYLAN A.B.D. *Died when H.E. bomb fell on site 2/4.*	18/10/40	AC 1	966290	RAF VR	30
MARTIN J. *Died of wounds. Bomb hit shelter, on day before, site 2/3. All 4 men dug out. Further 2 AFS personnel also died; no record here.*	27/10/40	AC 2	1051952	RAF VR	34
EWIN H. R. *No mention in 902 Operational Record.*	06/12/40	AC 1	840936	RAF AUX	-
VAUGHAN C. C. *No mention in 902 Operational Record.*	11/12/40	AC 1	840962	RAF AUX	40
TAYLOR J. W. *Killed by 3 H.E. bombs hits on site 2/5 (Flying Field) along with 2 persons of the Balloon Field Guard.*	15/03/41	AC 2	1021337	RAF VR	37
WARD E.W. *Bombs, incendiaries, and landmines dropped. Killed on site 2/22 in raid lasting from 0208 to 2016*	20/03/41	LAC	1195594	RAF VR	28

WEBB J. S. *Bombs, incendiaries, and landmines dropped. Killed on site 2/22 in raid lasting from 0208 to 2016*	20/03/41	AC 1	841347	RAF AUX	33
PHILLIPS W. J. *Died of cerebral haemorrhage in hospital*	23/02/42	CPL	840963	RAF AUX	-
ASHMAN A. H. *No mention in Operational Record.*	05/02/43	AC 1	840851	RAF AUX	37
MOODY E. *Fly bomb crashed on site 2/48.*	26/06/44	LAC	841049	RAF AUX	40
MANN R.H. *Billet badly damaged. Fly bomb on site 2/48.*	26/06/44	LAC	841284	RAF AUX	28
CROMPTON A. *Fly bomb on site 2/48. Recorded in 902 ORB as seriously injured this date. Listed died this date in BBRClub list.*	27/06/1944	LAC	1089069	RAF VR	38
PRIOR A. J. *Details not yet found*	03/02/4(?)	LAC	841267	RAF AUX	34

Squadron 903 Operations Record Book

HARDY F. T. *No mention in 903 ORB.*	26/08/40	CPL	841560	RAF AUX	40

COWNDEN A. E. 29/10/40
Defences reported his — AC 1 841926 RAF AUX 35
hut bombed.

MITCHAM A. A. 05/11/40
Defences reported
living hut bombed.
Died later of wounds. — AC 1 633772 RAF 20
Leave given to
brother 635505 to
visit him in hospital.

HILL J. H. 09/12/40
These Hills were
members of site 3/38,
possibly died from — AC 1 841766 RAF AUX 41
mine dropped day
before. Related?

HILL G. L. 09/12/40
These Hills were
members of site 3/38,
possibly died from — AC2 1177555 RAF VR 29
mine dropped day
before. Related?

GOLD R. D. 05/01/41
No mention in 903 — AC 2 1168166 RAF VR 33
ORB

COSSEY E. W. 12/01/41
1000lb UXB went off
at Platens Cottage,
by 903 HQ. There
had been ANO small — AC 2 1185146 RAF VR -
XB. Person next to
him had only
scratches! 60ft by
30ft deep crater.

BRACHER H. W.   12/01/41
Killed by bomb in
Herbert Road,
Plumstead. Full   AC 1   841819   RAF AUX   -
Military honours
given. May be late
report.

MILES A. R. *When*   17/04/41
active H.E. bomb fell
near crew. 4 injured.   CPL   841742   RAF AUX   29
He died in Lewisham
Hospital.

REBBECK J. B.   11/05/41
Died on leave when
H.E. bombs
demolished his block   AC 1   842019   RAF AUX   -
of flats at Welland
House, Peckham Rye.

Note: 901, 902 and 903 ORBs and their Appendices are available as downloads, at
http://discovery.nationalarchives.gov.uk/SearchUI/Details?uri=C2504858.
They each are found with end sequences in the series C2504858 - C2504863. The document archive number is given below.

NA Archive No.	Contents
AIR/27 2219	901 Full ORB day records
AIR/27 2220	901 and 910 Append to 901 and 910
AIR/27 2221	902 Day records from 11 Jan 1939
AIR/27 2222	902 Photo and Op. Orders 31May 1940
AIR/27 2223	903 Day records from 16 May 1938
AIR/27 2224	903 Append. and Op. Orders only

'Home Front' Memorials

The seed of thought that a specific war memorial should exist does not usually come into being by direct government action. A feeling for a need arises in a single or group of like minds who get it into an action path. Nelson's Column was built some four decades after his death. A group of 121 peers was its starting point to be created using public subscriptions. The Government provided the site.

There exist in Britain many WW2 memorials to *Home Front Service* personnel (Home Guard) and to *Home Front Civilian* personnel (civil services) but one dedicated to those who served in the *Home Front Active Forces* seems to be lacking.

Home Front Memorial

"The Home Front"
War Memorial at Bury, England

Home Front Service Personnel A procession in which different civilian occupations are represented including coalmining, engineering, carpentry. and munitions.

Royal British Legion, Stamford

"Hero's take on many forms, but anyone who served their country with pride and distinction is without a doubt a hero. May they never be forgotten. The whole nation owes a debt of gratitude for what they did."

"Lest We Forget"

Chapter 12 Memorials to Balloon Command

On War Memorials

Extracted from The War Memorial Trust web site:

'Why are war memorials important? A memorial is a physical object created to commemorate those who served in some way during a conflict. Generally, these memorials are erected by local communities associated with those remembered.

Each memorial is unique. They are important because they act as historical touchstones, linking the past to the present and enabling people to remember and respect the sacrifices made.

Memorials are an important source of information for young people in understanding the contributions and sacrifices made by past generations. They represent a focal point for remembrance, for both the individual and the collective, particularly on occasions such as Remembrance Sunday or anniversary events. The sacrifices made by so many for freedom need to be remembered. War memorials play a vital role in ensuring that is the case.'

Possibly the best-known piece of poetry on the need for military remembrance is:

'They shall grow not old, as we that are left grow old:
Age shall not weary them, nor the years condemn.
At the going down of the sun and in the morning
We will remember them...'

This is the fourth stanza from *For the Fallen,* written by Laurence Binyon in Cornwall, September 1914.

Whilst many war monuments commemorate lives given, they can also commemorate the hardship, loss of liberty, and bravery endured in giving service for humanity.

Some barrage balloon personnel served in dangerous active overseas war fronts. However, those at home so often also had dangerous active service brought right to them.

Those serving for No 1 Balloon Centre experienced considerable danger. Britain's worst time for balloon operations must have been from mid-June 1944. Then the landing places of the numerous V1 rockets, passing over every fifteen minutes, were too often on their locations, on and off the base.

The Doodlebug Poem
"Every night I lay in bed,
I hear strange noises overhead.
It's not the Angels in the sky,
But doodlebugs a passing by.
And as I watch their lighted tails,
My heart into my mouth nearly sails.
Will it, won't it, pass me by?
Stay up and doodle in the sky.
With apprehension I await
The cut-out of its hymn of hate.
The throbbing stops, the light goes out,
Enough to give the cat the gout,
Excitement tense on every face,
As under tables we all race.
Our shins we ski, our heads we bump,
And then we hear that awful crump -
So out we crawl like nervous wrecks,
And strain our ears to hear the rest,
Repeat the process jerk by jerk,
Then comes daylight - off to work.
Some fly east, some fly west,
Some fly over the cuckoo's nest,
And some don't!"
archive at www.bbrclub.org

Whilst personnel of the Barrage Balloon Command are rarely recognised as much as were the Battle of Britain pilots of Fighter Command and the crews of Bomber Command, balloon personnel also made major sacrifices and lived daily with the anxiety of being bombed, strafed, or attacked by fighter planes, and V1 and V2 rockets.

Amidst that situation they had to perform their arduous balloon duties regardless of the state of the weather. Over the duration of the war too many of these men and women gave their lives. Of those who served, two persons per hundred, did not live through it.

When Balloon Command was at its inception stage there were some senior RAF staff who felt the balloons should be referred to as 'sailing', not 'flying'. It did not take long for their contribution in the air to become recognised by those that knew of their importance as a major contribution to getting that war to its end. Acceptance as a worthy part of the RAF was well acknowledged when the war was over:

'When I was discharged, I was in the 'Battle of Britain Parade and March Past' at Buckingham Palace. The Balloon Barrage was in front of the Fighter Command. That is what the Fighter Command wanted, they wanted us in the front. Now I thought that was very good of them'
Stanley A Ross, 902 Squadron account, www.bbrclub.org.

This RAF Balloon Command came, and soon went, lasting only from 1938-1945.

The Balloon Barrage Reunion Club

The reunion club, BBRClub, was formed in 1945. Over the years since then, it has been building up a truly major knowledge source that is archived at the Club's website www.bbrclub.org. Most knowledgeable on the history of war-time balloon use is its webmaster, Peter Garwood (Club Secretary) who has been devoted to building this record as a free-to-all resource on balloon barrage information.

That archive site opens with this list:

Welcome to the Balloon Barrage Reunion Club

For all enquiries relating to this site please contact:

Peter Garwood: mobile 07412-869027 or email at
pgarwood@globalnet.co.uk

A section on seeking to find details of past RAF Balloon Command personnel is now available.

Click on the Barrage Balloon or click on the text underlined to navigate the site. These are examples from that collection.

 Home

 BBRC - All about us and our history and how to join.

 Aerial Bodies 2022 by Lily Ford

 The Rise of Women in Balloon Command

 Fallen Heroes of World War II- In Proud and Honoured Memory

The listings are now ordered by key word headings.

That freely available archive is a major monument to the Barrage Ballooners in the RAF. It contains accounts and photographs of men and women who operated, and otherwise supported, balloon operations. The information is comprehensive, including all of Britain.

National Memorial Set up by BBRC

It took a long time after the war ended for anyone to consider, and work at raising, a memorial to all the personnel of Balloon Command. The seed of this memorial must have been germinating for many years before it became a formally proposed idea.

It began as a survey, by Peter and Linda Garwood, of BBRClub membership around 2010. A formal proposal was submitted to the 19th of September 2012 BBRC AGM. It included a drawing based on a suggested design by Betty and Judith Tyson. Linda Garwood carried through the design aspect to completion. That statement went as follows:

'The general design is to have the elements of a Barrage Balloon, a balloon crew and a winch and cable engraved onto a stainless-steel plate. The plate is to be fixed to a slab of rough stone and angled in such a way that the balloon was visible against the reflection of the sky on the steel plate.

The idea is to have the upper part highly polished to reflect the sky, with the balloon together with the lower half non-shiny by dull etching. Based on the idea Betty and Judith Tyson, on the sides of the memorial there is to be additional explanatory information on the role of the Balloon Squadrons and perhaps the numbers of who lost their lives during the war.'

The meeting suggested the wording to be used on the top face:

'In Memory of the men and women who served with the Balloon Barrage Squadrons of the R.A.F Balloon Command during the Second World War.
Erected in 2013, by the Balloon Barrage Reunion Club'

A key question was where to site the memorial; somewhere appropriate and nationally relevant was needed.

Stemming from the aims of the Arlington Cemetery and the National Arboretum in Washington, USA, a national UK appeal was set up in 1997 for the nation to have its own *National Memorial Arboretum* NMA. Like so many memorials that have started up before, it began without money, without land, without staff and, not to forget them, without thousands of trees.

The site for the NMA is on a reclaimed gravel workings, gifted by Redland Aggregates, now Tarmac. Today it is far from a waste land, now being covered with trees and other beautiful settings ranging from the modest to the most magnificent of monumental architecture.

The place was initially developed by many volunteers, assisted with grants from the Forestry Commission and National Forest Company. A major funding boost came from the Millennium Commission. The NMA was officially opened in May 2001. Several appeals since have added facilities that cater for events of significance to be held there.

It is part of the Royal British Legion's, 150-acre visitor site on the edge of the National Forest in Staffordshire. Over 400 memorials nestle amongst 25,000 trees, Fig. 12.1. This was the logical site for a national Balloon Command memorial

Fig. 12.1 The NMA is the site of the BBRClub National memorial.

The first sketch, Fig. 12.2, of the club's national memorial was presented to the BBRC's September 2012 AGM, as part of the proposal statement.

Fig. 12.2 First sketch of a proposed National
Memorial to Balloon Command.

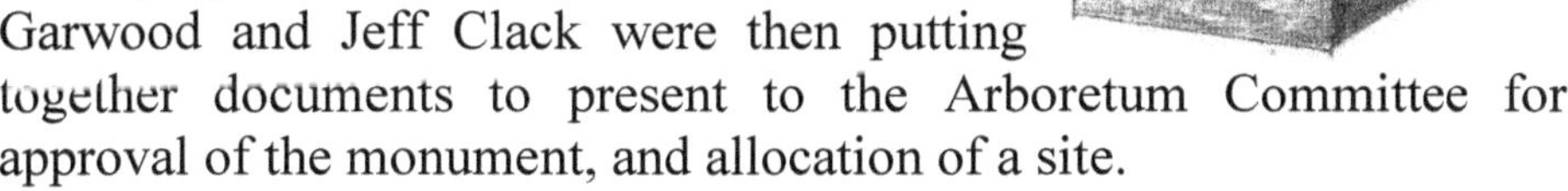

To build and site a memorial can be costly. During 2013 a BBRClub Memorial fund was created. The 2013 BBRClub Summer Newsletter reported that an improved drawing of the proposed memorial was available. Peter Garwood and Jeff Clack were then putting together documents to present to the Arboretum Committee for approval of the monument, and allocation of a site.

Sue McCall volunteered to carry the design aspect of the memorial. By March 2013 more of the design had emerged with the assistance of Pryorsign, who could make the image plate tableau.

Things so often come up against an unexpected factor. In June 2014 Peter Garwood learned that the NMA had changed its basic requirements for a memorial structure. From then on it had to be a block of stone, not a brick structure with attachments, and the lettering had to be engraved into that stone. The price would then be higher, but VAT could be claimed back in this case.

By September 2014 a price for manufacturing the memorial was obtained from Young Johnson Monumental Masons, being then at £6,800 plus VAT, for the complete supply and fix. They have had considerable experience with NMA memorials. The meeting accepted the quotation (presumably to be paid for from Club funds) with instructions to proceed 'as soon as was feasible'.

Additionally, to that cost, the NMA needed an application fee of £1,000, paid when the application was lodged. The price then was 'from' £7,800, depending in the number of letters to be engraved into

the stone. A block of blue granite was chosen for the memorial.

Time waits for no man, nor does it hold fixed prices for long! By mid-2015 a sort of quote had been obtained from Johnsons being in the range £8,000 – £9,000.

Fig. 12.3 Dedication of the National Memorial to the RAF Balloon Command.

The records then show that the NMA had agreed to a plaque being fixed to the top of the granite block. That plaque was in manufacture in September 2015 and the date of the dedication ceremony, Fig. 12.3, that was set for October to November 2015.

Formal dedication took place on 24 November 2015. A lengthy report of that event was sent out to members in the BBRClub 2014 Christmas Newsletter. A special colour printed booklet, *Veterans Memorial to Balloon Command Unveiled and Dedicated* was also sent to all members.

The memorial is located close to the main RAFA Memorial ('The Eagle') in the "RAF wood" area.

A solemn dedication service took place in the wooded area of the Arboretum. Peter Garwood officiated on behalf of the Club.

A detailed account of the ceremony, provided by Martin Postranecky, was published for members in the Christmas 2015 Newsletter and is recorded on the club website as *The Unveiling and Dedication of the Memorial to Balloon Command at the National Memorial Arboretum 24th November 2015.*

The ceremony was organised, in conjunction with NMA staff, by a Memorial sub-committee led by Jeff Clack, assisted by Peter Garwood, Marion Haill and Martin Postranecky.

The 'Order of Service', with its suitable readings and hymns, were all arranged for attendance and presentation by a padre, a bugler, a standard-bearer, and an organist.

A 'Visitor meet and greet' table was set up as the BBRClub stall inside the Pavilion Restaurant and reception building. Marion Haill and Lyn Clack welcomed visitors. Each person was given the 'Order of Service' booklet and a luncheon voucher. The booklet content is given in full in the Christmas 2015 Newsletter.

Some 30 persons attended the service, that commenced at noon and being held around the memorial. Peter Garwood led the event.
The pleasure of unveiling the memorial was given to Mrs Jodie Wood, daughter-in-law of the Hon. President Phyliss Wood, who sadly had passed away on the 16th of September 2015.

Three wreaths were laid at the foot of the memorial. The Act of Remembrance was read by the BBRClub Secretary, Peter Garwood, being followed by the bugler, Fig, 12.4, sounding the 'Last Post'.

Fig. 12.4 The Last Post being sounded.

The formal part of the event ended with a Royal Air Force Collect, the National Anthem, and the Blessing.

Special appreciation was recorded for the help provided by the local NMA Event Co-ordinator, Emma Cropper.

The only disappointment in this action must be that this memorial took too long to raise; a very few of the originally serving persons was able to attend.

The Imagery and Wording

The imagery, Fig. 12.5, shows a barrage balloon in action with its winch truck, bringing the original suggestion to life as the centre piece of the memorial. Linda Garwood, Fig.12.5, carried the design and execution to the final stage.

Fig. 12.5 National Memorial Plaque.

At the lower part of the top surface below the framed image, were these words, engraved into the granite and filled with gold lettering:

**In Memory of the men and women who served
with the Balloon Barrage Squadrons
of R.A.F. Balloon Command defending this country
and vital areas aboard during the Second World War
erected in 2015 by The Balloon Barrage Reunion Club - B.B.R.C**

The left side, from the front, bears these words:

**The purpose of the Balloon Barrage
was to prevent accurate bombing by
forcing enemy aircraft to fly at higher altitude.**

**This was achieved by steel cables held up by the balloons flown up
to 5,000 ft height. Aircraft which flew into the cables could be
seriously damaged or brought down.**

**The barrage balloons, filled with hydrogen gas, were flown either
from fixed or mobile winches, each operated by 10 men or 12
women, with 2 NCOs.**

On the RHS, from the front, is inscribed:

**R.A.F. Balloon Command was established in 1938
to provide additional air defence
of important targets and cities in Great Britain.**

**During the Second World War, by 1940,
There were over 1,400 Barrage Balloons in service.**

**By December 1942, some 10,000 men had been
released for other duties and replaced by over
15,000 W.A.A.F balloon operators.**

**In 1945 when the R.A.F Balloon Command was disbanded,
there were over 3,000 Barrage Balloons in service
with over 33,000 men and women serving.**

The memorial can be visited whenever the Arboretum is open to the public. It is advisable to first contact them to ensure a visit will be possible. Guides and motor trolley tours are available. There is a chapel there for private devotion.

No 1 Balloon Centre Kidbrooke Memorial Path

This section covers the decade long pathway, from 2019 onward, to endeavour to place memorials to the No 1 Balloon Centre on its original site. The complete aim was still not completed at the time of writing in late 2025 for it must await the completion of a new housing estate at the SE corner of the former RAF Kidbrooke site. Considerable effort has been expended by many people, especially by the students and staff of the Thomas Tallis school. Whilst much of their work has not been used yet, their contribution really had to be recorded. This is a summary of that path.
Peter Sydenham, Author and leader of the Kidbrooke Memorials Project.

Over the past decades since WW2, some of Peter's children, and grandchildren, were given a composition exercise to complete at their school. It was a recurrent teacher's theme; "What did your father (or grandfather) do during World War 2?"

Being interviewed many times led him to think more about his father, Henry's experience. It exposed the fact that many people really have little knowledge of life in this RAF service.

When Peter was a teenager in 1951, Henry told him he had been a Barrage Balloon operator stationed at the RAF No 1 Balloon Centre in Kidbrooke. Also, that he had sustained an injured by a V1 flying bomb that left his right leg injured; he could not walk very far and had to use a walking stick.

Repeating the tale became all too regular. He always asked them for a copy of what they wrote and to his surprise the older children began to know more about his experience than he did. Getting a long one-pager from a grandchild awakened Peter's interest. Perhaps Henry's story was of interest to his extended family? It was up to him to investigate it.

That led him to the Barrage Balloon Reunion Club BBRClub website. By around 2005 he had become engrossed in the topic, being then mildly obsessed, as one needs to be to keep up the chase for often elusive facts.

There was one clinching event that drove Peter to want to make sure the Balloon personnel's contributions were not allowed to be forgotten.

In 1950, whilst Henry and Peter were chipping wood for the frames of their self-build house in Adelaide, Henry casually told Peter that he was the only one in his hut, with ten people in it, to not die when it took a direct hit from a V1 bomb.

Not only did Peter want to know more about that incident, but also could he identify any of the men or women he had served there with? What was it like to be a balloon operator? Was it an easy task, without danger? When did he get injured - and more?

After getting Henry's Service Record, his Medical Record, and the Squadron Operational Records Books ORB, things began to unfold. By 2012 Peter's version of his personal account was published on the website, www.BBRClub.org as: *'840931 Cpl Henry Sydenham RAF No. 1 Balloon Centre (901, 902 Squadrons)'* by Peter Sydenham. To access this, use this search title in the list of articles given on that website.

Preparing that article made Peter realise that Henry may well have had that kind of guilt that occurs in such cases - why was he the only one to survive? He would have surely felt the need to honour those who died. With Henry's passing in the 1960s that had become Peter's task.

Having prepared Henry's own story, it became clear there was much more to research to better understand life on the base that became the home of No1 Balloon Centre in 1938.

It also raised the need to get memorials in place for Peter appears to be the last child contactable of all of those who served there. If he did not provide some kind of record for all to find and see, these memories will pass on forgotten.

An account, in three parts, was published in BBRC Newsletters. The consolidated article was then posted onto the BBRClub website as:

No 1 Balloon Centre, Balloon Command, RAF Kidbrooke 1938-1945. Squadrons 901, 902 and 903, by Sydenham P H

But can one work up the key issues of raising a physical memorial from so far way in Adelaide where the author lives, and with no local knowledge at all the Kidbrooke area? With the Internet and email the distance factor was made possible, but not without difficulties. To make it harder, from the start no person living could be found who would be his local co-researcher; that has been a sorely missing link.

In February 2013 the opportunity a arose to see the Kidbrooke area first-hand. The memorial project had no connection to local people, so it was starting from a very basic place. The initial target was to identify a suitable site and find a host organisation who might support the local management needed along the path to a memorial's dedications.

It was a problematic flight from Adelaide to London. At the airport stop in Singapore the scheduled time of departure onward was delayed by 6 hours! A hastily made on-line booking at Singapore booked a car to pick Peter up in the Heathrow airport, making it possible to still arrive in Kidbrooke in time to link up with the BBRClub representatives and allotment members at the Kidbrooke Park allotment site.

It was a clear, but very cold frosty February morning, a kind of day the barrage ballooners would have dreaded for their cables, ropes and the equipment would have all been covered with a thin sheet of ice that had to be cleared off as they went about their duties. On that day of Peter's visit his large suitcase had to be lugged over many frost-hard paths - with severe jet lag in evidence for him!

Fig. 12.6 Common shed for allotment members.

Whilst there at the allotment he wanted to locate the area in which the V1 bomb had apparently landed, learn more about the allotments and its membership, and investigate where a memorial could possibly be sited in the local area. Attending were Catherine Clancy and Bob Tyrer of the Allotments; Marion Haill and Martin Postranecky of the BBRClub and Peter Sydenham. The allotment site has a common shed for storage and meetings, Fig. 12.6.

Using the western Kid Brook ditch, seen behind the metal chain-mail fence, and the existing narrow (modern) bitumen road, it appeared to be that the V1's impact site was centred around plot 21; easily found being near to the white spotted, blue painted shed, Fig. 12.7.

Fig. 12.7 Expected V1 landing place.

A laminated A3 paper memorial, Fig.12.8, was placed, without ceremony, inside the day hut window. To know more about this RAF site the records of the National Archive NA at Kew, the Imperial War Museum, the London Museum, the London County Council archives, and Royal Air Force records, still need more study.

An archaeological dig, including a metal detector scan of the site in the Kidbrooke Park Allotments might reveal more artefacts of that V1 strike.

Fig. 12.8 Initial Memorial to the Fallen, February 2014.

Allotment users are encouraged to collect items with their locations and note them, as they work their plots. It might help find out more about these tragic events.

At the time of that visit Peter had a preconceived brain-hit idea of a suitable memorial site as the starting point of his search for a suitable place. Looking over the fence to the NE of the allotments was then seen a cracked concreted ground, full of weeds. That might just be a good spot for a small green park in which a memorial plaque could sit! (See later for an amazing, fortuitous, coincidence.)

Enquiries were made of the Greenwich (Borough) Build plans for development in that location. The enquiry, however, kept being deflected into the then emerging massive Kidbrooke Village works organisation. It was too early to expect a site to be chosen.

However, at the time of commencing writing this book, in May 2022, it had become clear that area then seen by Peter on his 2014 visit was to become a housing estate.

A suitable site seemed to have had been found, with help from Greenwich Build. However, that estate was still in the planning approval drawing stage – see below for details.

Peter returned to Australia in March 2014, being most unsure of where to try to locate any memorial. However, he had then learned that the Thomas Tallis school was situated on land formerly used by RAF Kidbrooke. Using Google maps, he was able to find a suitable spot near to the school's main entrance. The school had been rebuilt in its current location, from the location where the new estate is being built.

Fig. 12.9 Suggested location of the memorial, taken March 2019.

Peter, not being in the UK did not help things along, but Marion Haill and Martin Postranecky of the BBRClub committee were able to visit the area to see if the site he had seen was suitable. In March 2019 Martin found the spot the earlier Google Earth search had found, Fig. 12.9. The location between the two small trees was ideal, having grass for meetings; being behind the protective fence; and yet easily seen from the foot path.

A local helper was needed to keep things moving. In January 2019 the project got a boost when John King, a local resident, and keen local historian, contacted Peter via Peter Garwood. He had learned, from Peters' article posted on the BBRClub website, of the interest in having an history plaque put up in Kidbrooke in memory of the No 1 Balloon Centre's presence at the then RAF Kidbrooke site. A spot in the school grounds seemed to be a good choice.

During late 2019 informal contact was made with School staff over that suggestion; but it was not progressing. From early January 2000, the Covid Corona virus had just started to impact on the world's ability to get things done. The memorial project was losing impetus.

Just what kind of memorial to go for was unclear. John King made enquiries of Greenwich Borough councillors who showed interest, but made no suggestions on what to do. John, however, also emailed me:

'The good and more positive news is that the Royal Aeronautical Society has informally agreed to embrace your plaque in its scheme of historical plaques.'

That RAeS Heritage scheme is prestigious and well publicised. What a stroke of luck? Never really being sure about raising a significant memorial, this was a real fillip for the project.

There was no deadline for this to happen, that allowing more development of the full memorial design to be pursued. The requirement that the memorial is situated where the public would see it, was added as a condition of the RAeS heritage award.

Knowing that making the RAeS bronze plaque casting would cost around £1200 it was time to tie down some support from the BBRClub toward the memorial. Peter had previously submitted a 'vision' proposal, seeking a reaction from members at the BBRClub Informal Meeting of 12 September 2018. It was well received, with £2500 being ring-fenced to the building of such a memorial - if it goes ahead. This affirmation had added confidence to continue the project.

Just how it would be pursued was up to Peter as there possibly would be no more financial support from the club due to dwindling club membership of persons who served there in WW2.

Developing a sound application for the RAeS award took over 18 months. Covid restrictions, including school lockdowns, did not help.

The best location for it was then seen to be a site within the school grounds. The Business Director, Maggie Shields, of the Thomas Tallis school was contacted by email and a face-to-face meeting was set up at the school on 2 July 2019.

In attendance were Maggie Shields, Marion Haill, Martin Postranecky and John King. The case was made for a street side monument. It was also suggested that the design be part of a school's art project. The meeting appeared to have strong support for the idea. Maggie took our proposal to the school officials.

However, Maggie left the school shortly after that, and the link needed to be revived in May 2020, the contact then becoming with Cheryl Campbell, the new Business Director. By mid-May communications by email were again working.

On 17 July 2020 Cheryl rang John King to advise that the school was embracing the idea. However, the commercial partner, who owned the school, had to give its approval to site it there. More detailed was needed for their consideration. The RAeS application, however, could not be lodged until this location issue was resolved. It all took time!

Cheryl replied on 3 November 2020:

'Good news, the Local Authority and the funders are both on board with the plan for a memorial. The funders have also offered to fund a prize for a design competition'.

Cheryl had been working hard for this cause. Many people had been consulted including Jules Rutt, CIWFM, Head of Contracts & Property Management, Directorate of Regeneration Enterprise & Skills and the Royal Borough of Greenwich.

Emma Cannings, of Equitix, replied that their SPV Board (Special Purpose Vehicle arrangement) was of the opinion:

'I am pleased to confirm that the SPV Board are more than happy to support this idea, I will need to run it past G4S [their site security agency] but I do not envisage that being a problem, we will need to speak to them about how the area is maintained etc. but that shouldn't be prohibitive.'

With these recommendations to hand, a specific school site was selected on 27 July 2021. It is close to the spot originally suggested, but one that is more visible, and more secure. It was marked up by a small group, Fig. 12.10. Present were Kristina Ferris, Head of Graphics and Deputy Curriculum Leader: Design Technology; Cheryl Campbell, Business Director; John King the local representative, and Richard the technician.

Fig. 12.10 Site being marked up on 27 November 2021.

At the meeting Kristina announced there was to be a school design competition for the memorial, ending on 1 April 2022. The design brief issued by her is shown in Fig. 12.11 below.

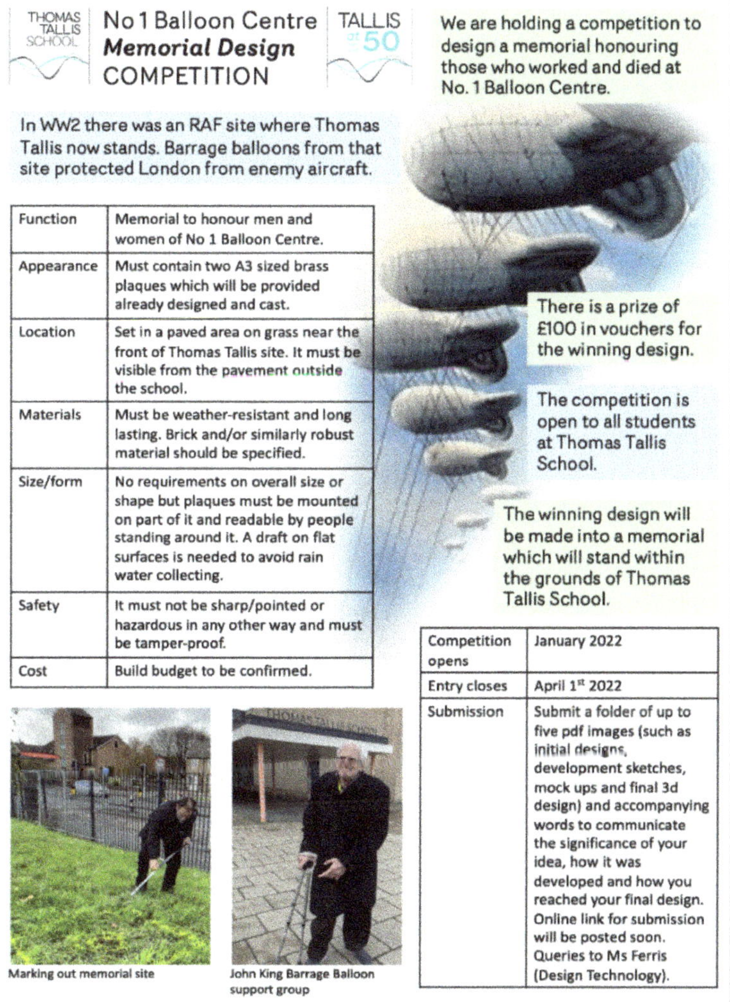

Function	Memorial to honour men and women of No 1 Balloon Centre.
Appearance	Must contain two A3 sized brass plaques which will be provided already designed and cast.
Location	Set in a paved area on grass near the front of Thomas Tallis site. It must be visible from the pavement outside the school.
Materials	Must be weather-resistant and long lasting. Brick and/or similarly robust material should be specified.
Size/form	No requirements on overall size or shape but plaques must be mounted on part of it and readable by people standing around it. A draft on flat surfaces is needed to avoid rain water collecting.
Safety	It must not be sharp/pointed or hazardous in any other way and must be tamper-proof.
Cost	Build budget to be confirmed.

Competition opens	January 2022
Entry closes	April 1st 2022
Submission	Submit a folder of up to five pdf images (such as initial designs, development sketches, mock ups and final 3d design) and accompanying words to communicate the significance of your idea, how it was developed and how you reached your final design. Online link for submission will be posted soon. Queries to Ms Ferris (Design Technology).

Fig. 12.11 Brief issued for the design competition.

With a site selected and approved, the RAeS application was formally submitted on 9 April 2022, with a small alteration to the wording. It had been through the RAeS first step, the *Scrutiny Panel,* getting unanimous support. On 28 May 2022 the application was moving through the *Medals and Awards Committee*, and then on to the RAeS Council for final approval. It was approved in that form in July 2022.

Back to progress of a memorial on the school site. Submitted designs were exhibited at the school's 50th Celebrations held on 16 July 2022. A welcome was included for Kitty and Beth James, third cousins of Peter, who represented the Sydenham family, Figs. 12.12 and 12.13.

Fig. 12.12 LHS of exhibition of designs.

Fig. 12.13 RHS of the exhibition of designs.

The exhibition of submissions was open in July 2022. Those attending the exhibition were asked to indicate the best design in their opinion, by adding a dot sticker. The Reuben Elliss-Wood design was voted the best, by far. His design being Fig. 12.14.

Reuben's explanation expressed the thinking behind his creation:

"For this memorial I wanted to shed some light on the people who worked and lost their lives in the Barrage Balloon Centre in World War Two. I decided to incorporate the modern ideas of the 'Tallis Habits', Fig. 12.15, into the memorial. The colours make it a bright and friendly monument even though it is about dark times. I would also love to see my own model in the Tallis grounds for people to be educated about the history of the area".

Fig. 12.14 The Reuben Elliss-Wood design.

Fig. 12.15 Tallis 'Habits of Mind'.

The practical problem, however, was that making a low-cost memorial that would retain its brilliance and robustness in an external setting was not feasible with then expected BBRC grant.

For six months after the exhibition ended (still in the Covid period) numerous attempts by email were made to purchase a simple 'plinth' on which to mount two A3 sized brass plaques.

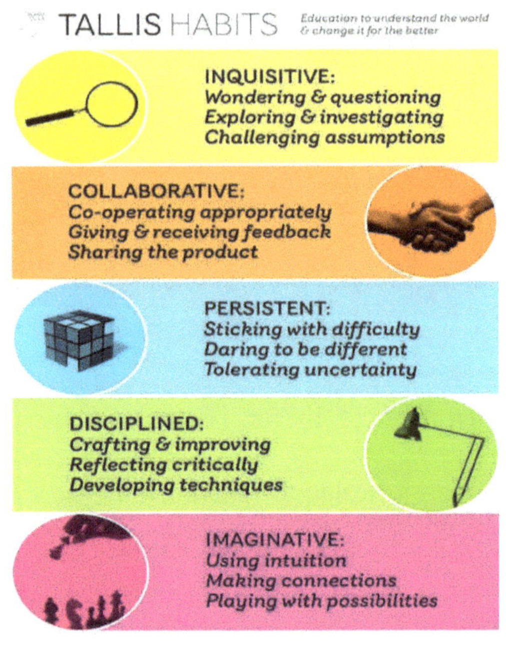

Amazingly no progress could be made. Suppliers just did not reply to dozens of email requests for simple objects like a stone or fabricated metal stand to carry the two brass plaques. More on this memorial School opportunity is covered later.

RAeS Heritage Plaque is Placed

This had been approved in July 2022. Then came a major halt to more progress. The BBRClub could no longer provide any more than half of the cost of the purchase of the RAeS bronze plaque, Fig. 12.16.

Fig. 12.16 RAeS plaque.

In mid-2023, with much regret, the school was advised that their outside memorial location could not be followed up by Peter. It was suggested that Reuben's design, too good to not use, might be made by future school art students as a monument for an inside location. That creation could use commonly available indoor long-lasting materials and was well suited to the school's art student skills and syllabus.

At the time of writing, it was not known if that realisation will be followed up by the art branch of the school. Peter sincerely hopes so!

Reuben's design is not so much a memorial to the fallen ballooners carrying out their roles in the context of Central London but is more the school's recognition of the Barrage Balloon activity, expressed in school 'habit' colours.

With cessation of that school site plan another site was needed to erect the RAeS plaque in public view, and to also place another plaque to the fallen, the original intention.

Over the latter part of 2023 a site for the RAeS plaque was located.

Over the past nine decades the land of the former RAF Kidbrooke base has gradually been built on – reused for a storage company, the allotments, the Thomas Tallis school, and today for a sizeable estate under development called *'Land to the West of Kidbrooke, Park Road, Kidbrooke SE3 9PX'*.

The nearby *Depot* public house was recently built on the site of what was, during WW2, a place for storing barrage balloons, Fig. 12.17. It now stands there - hence the pub's name.

Fig. 12.17 The Depot Public House in Kidbrooke.

Contact was made with Young's Brewery's head office asking if they would consider placing the RAeS plaque on the inside wall of the restaurant hall in the Depot public house, that with a wall already decorated with barrage balloons. They were enthusiastic with the request to host the plaque. It was installed ready for the unveiling ceremony held on 29th April 2025, Fig. 2.18.

Fig. 12.18 Plaque in place at the Depot public house. RAeS.

Those in attendance in Fig.12.18, were (L-R) RAeS CEO, David Edwards, Gp Capt. Mike Hawkins RAF Ret'd of the RAeS Heritage Scrutiny Panel, and Past-President, Air Cdre Bill Tyack RAF Ret'd, standing alongside the newly installed. Representatives of the Sydenham family and the Barrage Balloon Reunion Club were unable to attend the event.

Memorial to the Fallen

Another plaque, under development in August 2025, reflects the original requirement of honouring the Fallen, shown in their operational context. A rough draft of the casting is given in Fig. 12.18. The imagery will be of much improved quality.

Fig. 12.19 Memorial to the Fallen.

Regarding a possible location, a key build planning document had been released:

'Kidbrooke Housing estate London 5 April 2022 report. Ref No 22/0001/F.'

Following into Item 6.1 in that planning report, is 'Figure 3 Proposed ground floor plan'. The central area, designated as a community space, is planned to be a small green area.

Amazingly, that is the very area where I had looked, in February 2019, over the allotment fence to see a discarded concreted area, overgrown with weeds. I immediately envisioned such a place for Kidbrooke residents to use, see the red dot in Fig. 12.19.

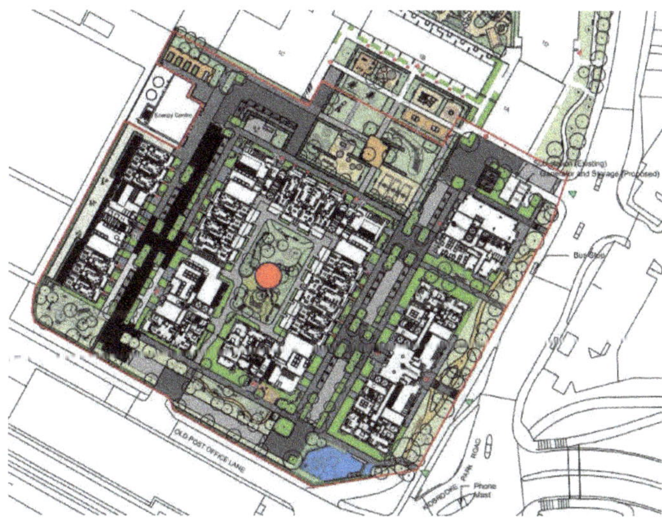

Fig. 12.20 The central park (red dot) suitable for the 'Fallen' memorial.

Contact was made with the design's sign-off person, Giulia Acuto, asking if they would consider placing the above Memorial to the Fallen there.

That request was forwarded to Lee Christie, of Greenwich Builds, the Engagement and Consultation Manager, Royal Borough of Greenwich (Lee.Christie@royalgreenwich.gov.uk, who has shown encouraging interest in the idea. His reply, on 18 May 2023 was:

'Hi Peter,
I've been passed your email through Planning, where I am the Consultation and Engagement Manager for Greenwich Builds, delivering the new homes in Kidbrooke Park Road, on the old Balloon centre. Your website (www.midhurstmemoirs.com) is very insightful and I read your request with great interest, where we are very happy to look into this.
Much of the detailed design work has yet to be started on the communal spaces on our Kidbrooke Road sites, so it may take a while, but I will make sure that you are kept up to date.'

And more recently on 6 January 2024:

'I am sorry that I cannot give a definitive answer sadly, as the 'estate' has not been named yet. Construction is in its early stages on Kidbrooke Park South and the whole development will not be complete until 2026. I have forwarded your email to the Council street naming officer and welcome any suggestions you have for block naming, with your local knowledge.

Being at the early planning stages of the estate build, this suggestion can only be kept at the ready for when that green community area is being detailed. This could be well into 2026!'

And again, from Lee Christie, on 16 July 2025:

'Phase one is complete, and phase two will start handing over next year, where the landscaping will be the final element.'

It is looking good - but patience is essential here. The task of getting them all in place will possibly need to pass on Peter's children!

With the two plaques set up decided and this book published online, recognition of the role played by members of the No 1 Balloon Centre in the protection of central London from bombing will have been completed.

Perhaps, also, there will be an indoor memorial built by students at the school.

Other Barrage Ballooner Recognition
Previously mentioned are the Training Centre shield, that recognised the WAAFs at Cardington - see Chap 7, Fig. 7.3, and the wall decoration of barrage balloons at the Depot pub, by Kidbrooke railway station – see Chapter 10, Fig, 10.5.

Tribute in the *Sunday Times*.
At my visit to the Allotments in 2019, allotment plot holder Bob Tyrer, was present.

He subsequently wrote one of his humorous wine columns for the *Sunday Times,* titled '*On the Bottle'.*

Bob was seeking to grow and make his own wine from vines he has cultivating on his allotment in Kidbrooke, situated nearby where the hut that was demolished once stood.

Inspired by this piece of local history he wrote the following tribute article appearing in the 23 February 2014 issue p51:

THE DISH

On the bottle
The secret history of my allotment

BOB TYRER

THREE OF THE BEST

LANGHORNE CREEK MARSANNE ROUSSANNE 2011
Delivers orchard fruits in abundance. Open it for pure pleasure and drift away. (*£9.99, Marks & Spencer*)

MY text for today is "Balloon Barrage Reunion Club", which turns out to have a big role in my efforts to create Château Bob in the wilds of southeast London. Its web address, bbrclub.org, sounds like a famous St James's wine company; and there's a whiff of espionage to the story.

My little allotment vineyard, it turns out, occupies a corner of the former RAF Kidbrooke, perhaps the least-known air force station of the Second World War, hidden among

Second World War, hidden among tracts of pebble-dashed semis and sports grounds between the A2 and the A20, the great trunk roads in our part of the world. Ostensibly, it repaired barrage balloons — the huge inflatables that floated above London hooked to earth by steel cables to hamper incoming bombers. Secretly, balloon repairs were a front for something else.

Peter Sydenham, who as a south London boy emigrated to Australia with his family in 1961, wrote from Adelaide to the allotment committee about his father, the only survivor of a flying bomb strike on RAF Kidbrooke, which killed nine people. Peter wanted to come over and erect a monument to them. Early this month, with Peter and two members of the BBRC, I stared at the weeds and old bricks where — judging from aerial photos — he thinks the hut stood. It's about 20 yards from my vines. The reunionists have added piquancy by revealing that Kidbrooke was an outpost of the Bletchley Park codebreaking operation and that counter-intelligence agents were taught Russian there during the early Cold War.

Somewhat inadequately, here are my own tributes to this history: a bottle with a balloon on the label, a lovely wine made by a Frenchman who huffs and puffs so much he could fill a dirigible, and a cheap but good champagne that will make you float ∎

DOMAINE DES
EYSSARDS
BERGERAC ROUGE
2011
Pascal Cuissard rarely stops talking but finds time to make this hugely pleasurable bargain. Like him, it needs air, so open a day ahead.
(£7.99, *Waitrose*)

CHAMPAGNE
VEUVE MONSIGNY
NO 3
It, too, needs air to shunt aside its initial blandness, revealing lithe muscularity and food-friendly flavours. Not an aperitif style.
(£12.99, *Aldi*)

Epilogue

Few of the children of serving persons at No 1 Balloon Centre have been identified, who may be still with us at this time.

Brenda Parson (nee Happs). Her father, Cpl Bill Happs, was one of the persons who died by the V1 direct hit of 18 June 1944. Evidence shows she was still with this world when she reported her annual Fell Walk on 7 January 2024. See: https://www.wainwrightwalking.co.uk/angletarn-pikes-and-brock-crags/   Her father's story is covered in Chapter 9.

Stanley Ross   Stanley was one of the three *Messengers* from the Royal Exchange in London, who enlisted with Henry Sydenham in the same queue. Stanley started out in Squadron 902 but moved from No 1 Centre soon after being posted there. He returned to the Royal Exchange after the war as the Manager of their dining room.

He was the father of Vera Pullin who moved overseas after the war. She provided information, much being used in this book, on his time starting in the Auxiliary Air Force in 1938, and then being embodied.

Vera used to be the respondent for his family, but now brief messages come from her husband, Arthur - still using pullinavj@gmail.com The last message was not answered.

Peter Sydenham His father was Henry Sydenham whose wartime story has been well covered in this book and cited articles.

After the war he still had trouble with his wounded leg, but he pushed on with his life. There were no post-war opportunities as a messenger at the Royal Exchange because advancing communication technology had much reduced the need for their kind of contributions.

Henry was offered a post-war training scheme to learn a trade. He worked in London with Dorman and Long, a structural steel company. Using night school, he became a skilled structural steel draftsman. Peter saw little of Henry for he left for work each day at

7am and returned home, after evening classes somewhere, after 8pm. He continued to suffer badly with asthma.

By 1948 it was apparent life for this Sydenham family of three was making little financial progress in post-war England; add to that the travel to work on public transport and the, not the least, bad weather! He and Vi were easily persuaded to make a new life as sponsored migrants to Adelaide, South Australia.

That was an easy decision for them to make. His sisters, Jinnie and Fran, had married WW1 Australian soldiers after WW1 and went to live in Adelaide. They often sent letters to Henry in the 'old dart', on how good life was in Adelaide – exotic fruit in abundance from home gardens, sun galore, close by beautiful beaches, ability to own a car - and most of all, being able to own land freehold and build one's own home on it.

They left as '10 pound Poms' on the SS Mooltan, to arrive in Adelaide in late October 1951, with their allowed 1 cubic yard crate to bring with their meagre possessions.

Life turned out as promised. Within a few weeks Henry bought a nice sized suburban building plot on which he, and Peter built their own 3-bedroom house on it - with their own hands. The car and the house were all paid for in just two years. Even I had a car, of sorts, of my own.

Henry did well as a structural steel draftsman and was promoted to a senior position in Forward Johns and Waygood, an Adelaide steel fabrication factory. He was involved with many commercial builds in the city of Adelaide, and in the structure created to launch Blue Streak rockets at the Woomera test site.

His life had been hard, being brought up pre-WW2 in a large poor, inner London location; with chronic asthma at a time when there were no satisfactory treatments: being a heavy cigarette smoker; enduring that WW2 stress of sailing balloons for all days, in all weathers: hard work learning as trade in evening school; and then designing and physically building our house. That all took its toll on Henry; his siblings had inherited heart problems as he had.

In March 1961 Henry died of a second heart attack. At that time heart problems had few effective treatments. If you had a heart attack and survived that one, it usually meant another more severe one would come within months to take your life.

He had enjoyed the fruits of his labour in Adelaide for just less than a decade; but they were the best in his and Vi's life.

Henry's (often called Harry or just Syd) grave at St George's Church, Magill, South Australia.

Appreciations and Acknowledgements

Many people have assisted here. They include:

Peter Garwood, BBRClub Secretary, with special mention
Martin Postranecky, BBRClub Newsletter Editor
Marion Haill, BBRClub Treasurer
Vera Pullin, BBRClub member
Arthur Pullin, BBRClub member
Members the 'Airfield Information Exchange' website
Percy Featherstone, one time stationed at RAF Kidbrooke
Katherine Clancy, Kidbrooke Park Allotment Association
David Wise, Kidbrooke Park Allotment Association
Bob Tyrer, Wine Writer, and Kidbrooke Park Allotment Association
Andrew Dawson, assistance with files at the National Archive
Pete Gourri, Kidbrooke resident
Maggie Shields and Cheryl Campbell, Business Directors, Thomas Tallis School
Kristina Ferris, Head of Graphics, Thomas Tallis school
Tallis School pupils and associated teachers and technicians who became involved in the memorial design
John King, a special thank you for the breakthrough leading to the RAeS award
Scott Phillips, Royal Aeronautical Society RAeS
Gp Capt. Michael Hawkins, Royal Aeronautical Society RAeS
Jules Rutt, Royal Borough of Greenwich
Emma Cannings, Equitix
Emi, Manager, The Depot Pub, Kidbrooke
Lee Christie, Greenwich Builds Engagement and Consultation
The James Family, Cambridge

Image attributions are given in the Illustrations list.

With sincere thanks.
Peter H Sydenham, Adelaide, 15 September 2025

Born in London in 1937, Peter was evacuated to Midhurst in West Sussex. With his family, he migrated to Adelaide, South Australia in 1951, where he now lives. His working life began in Australia as an electrical trades apprentice, merging into a career in Engineering from 1961. Upon gaining the BE (Hons) and ME at the University of Adelaide, with his wife and baby daughter he went to Warwick University for a PhD research period in the 1960s. After a decade in geophysics at the University of New England, NSW, he returned to Adelaide as a Professor of Electronic Engineering. In 1986 he was awarded the DSc in Engineering by the University of Warwick, UK. He retired in 1998 to follow writing and handicrafts. He and Pat spent five pleasant years in the Cotswolds.

Peter has always been attracted to non-fiction writing. His trade work led to articles in the magazine, *Electronics Today International.*

During his 35year academic career, he has also authored or edited, over 20 text and research books, and the usual 100 plus scholarly papers. He was a book-series Editor for John Wiley, in Chichester. Since 2014 he had been researching, writing and publishing a 4-book series on his WW2 time experiences in Midhurst.

Writing far away in Adelaide, Australia, has not been a problem; the Internet, and its email, provide most of what is needed today.

His interests have covered numerous handcrafts - with mixed successes. He completely rebuilt and modified an Amilcar racing car in 1954. His own designed 1958 electronic base-guitar was a project to forget, but his several house building extensions were well done. His 2017 classic decorated wrought iron balustrades received unsolicited craftsman acclaim (but wore out his shoulders)! He has constantly been building onto their homes to provide for their 6 children - who now live in Adelaide, Dubai, and Melbourne. Pat and Peter now have 12 grandchildren.

Illustrations

No 1 Balloon Centre, Kidbrooke
with sources

Fig. 1.1 Balloons flying around London.
   Google
Fig. 1.2 Coloured etching of the Montgolfier hot air balloon.
   Google
Fig. 1.3 British Observer Balloon of WW1.
   Flikr Anders
Fig. 1.4. Parts of a British Mk VII Balloon.
   Flikr Derek66
Fig. 1.5 Life was easier at times.
   Bing image

Fig. 2.1 Balloon Command badge.
   rafweb.org/badges
Fig. 2.2 Balloons on their winches ready to "let up" at Cardington.
   Painting by Grace Kingston.
Fig. 2.3 Recruitment poster.
   IMW
Fig. 2.4 Air Vice-Marshal O T Boyd.
   Google
Fig. 2.5 WAAFs on parade with balloons.
   https://ww2db.com/image.php?image_id=31338

Fig. 3.1 Appliances for examination of fabrics and cloth.
   See text
Fig. 3.2 Students at a Joint Services School for Linguistics.
   militaryintelligencemuseum.org
Fig. 3.3 RAF Hampden bomber.
   http://hameldownboys.com/
Fig. 3.4 Air Minister sees barrage balloons,1938.
   Pathe Film no 979.49